T0336510

# Fourier Transform and Its Applications Using Microsoft EXCEL® (Second Edition)

Online at: https://doi.org/10.1088/978-0-7503-6044-9

# Fourier Transform and Its Applications Using Microsoft EXCEL® (Second Edition)

**Shinil Cho**

*La Roche University, Pittsburgh, Pennsylvania, USA*

**IOP** Publishing, Bristol, UK

ISBN    978-0-7503-6044-9 (ebook)
ISBN    978-0-7503-6042-5 (print)
ISBN    978-0-7503-6045-6 (myPrint)
ISBN    978-0-7503-6043-2 (mobi)

DOI    10.1088/978-0-7503-6044-9

Version: 20231201

IOP ebooks

British Library Cataloguing-in-Publication Data: A catalogue record for this book is available from the British Library.

Published by IOP Publishing, wholly owned by The Institute of Physics, London

IOP Publishing, No.2 The Distillery, Glassfields, Avon Street, Bristol, BS2 0GR, UK

US Office: IOP Publishing, Inc., 190 North Independence Mall West, Suite 601, Philadelphia, PA 19106, USA

*To my family, Mari, Felix, and Michael*

# Contents

# Preface

## Preface to the first edition

The Fourier transform (FT) is one of the most useful mathematics tools and it has been applied to various physics and engineering applications. There are many textbooks on FT focusing on its mathematical theory. However, in order to acquire the knowledge and skill to conduct FT analysis, students of science and engineering need to practice realistic examples through which they can confirm and understand FT theories.

This book demonstrates FT examples selected from physics to illustrate the practical computational issues on FT and to develop a know-how for obtaining fine spectral analysis. Rather than presenting rigorous mathematics, this book offers examples for readers to try and to use Fourier transform for themselves.

With the recent remarkable advancement of computers, it is possible to study this subject with a single personal computer even at home. Microsoft EXCEL® has the fast FT routine in its built-in analysis tool pack and requires virtually no additional programming. This is a great advantage in acquiring the essence of FT step-by-step while also applying it to physics phenomena. This book uses only EXCEL for all analyses and graphs so that readers can acquire the spectra on their PC screens by themselves. Equations and parameters for EXCEL—based an acoustic FT analysis —are also included for reference purposes.

This book also discusses a method for computing power spectral density that is an alternative to the FT. It estimates spectral density from time-series data without FT. Both methods are compared using the same data set to learn more than could be investigated with one method alone.

Originally, the acoustic spectral analysis described in this book was created for the first STEM project at La Roche College, and extended to cover more FT-related subjects. The author hopes readers will gain the knowledge and skill of Fourier transform through this book, and step into more advanced subjects.

S I Cho
chos1@laroche.edu

## Preface to the second edition

In this second edition, I have updated some inaccurate descriptions from the first edition, and added more applications of FT.

The second edition consists of two parts. The first part consists of chapters 1 to 3. Chapters 1 and 2 provides a foundation of FT, which can be applied to chapter 3 to conduct an acoustic spectral analysis. In chapter 2, the algorithms of and one- and two-dimensional fast Fourier transform (FFT) are added. Their VBA codes are given in the appendix. The second part contains chapters 4 to 6, where more

advanced topics are described including linear response theory, Fourier optics and stochastic process, spectral methods for solving diffusion equations, Green's functions, and quantum Fourier transform for Shor's algorithm and quantum walks. They are concise introductions to these advanced subjects. The appendix contains notes for EXCEL and the VBA codes along with detail mathematical calculations so that readers are able to confirm their calculations.

Late Summer 2023.
Shinil Cho
chos1@laroche.edu

# Acknowledgements

The author acknowledges valuable feedback from readers of the first edition. Dr Shoichi Kiyomoto, CEO of ASD, Inc. in Tokyo suggested valuable topics on Fourier transform. The author is grateful to Dr John Navas, Senior Commissioning Manager, of IOP Publishing who encouraged me to write the second edition, and Ms Phoebe Hooper and her staff for brushing up the text and careful proofreading. Without their help, this book would not be published.

# Author biography

## Shinil Cho

 Shinil Cho attended Rikkyo University in Tokyo, Japan for his BS degree; Seoul National University in Seoul Korea for his MS; and the Ohio State University in Ohio, USA for his PhD. He held postdoctoral fellowships at the Ohio State University and University of Florida, and he was also a visiting professor at University of South Carolina. He has been at La Roche University since 1995. Currently he is a Professor of the Department of Physics at La Roche in Pittsburgh, Pennsylvania.

He has conducted research in cryogenic magnetic resonance spectroscopy below 1 K and biometric fingerprint authentication. His current research interest includes quantum computation, biometrics, and physics education. Other than physics topics, he has many publications and has done many presentations in London, Tokyo, Hongkong, Singapore, and several cities in the United States.

# Part 1

Foundation and Hands-on experiment of
Fourier transform

**IOP** Publishing

# Fourier Transform and Its Applications Using Microsoft EXCEL®
## (Second Edition)

**Shinil Cho**

---

# Chapter 1

# The principle of superposition and the Fourier series

There are numerous periodic motions—such as oscillations and waves—observed in nature. Differential equations describe their periodic motions in the time domain. They can also be analyzed in the frequency domain in order to acquire distributions of frequency components and their amplitudes because arbitrary oscillation patterns can be constructed with sinusoidal functions using the superposition principle. Such sets of sinusoidal functions are called the Fourier series, originating from a thermal conduction problem solved by Joseph Fourier.

In the first chapter of this book, we show how oscillations on a string and a membrane, and diffusion of heat and particles can be constructed using a set of possible solutions of their differential equations to introduce the Fourier series. We also describe how to construct a Fourier series of a given periodic function, and show examples of periodic functions in terms of their Fourier series.

## 1.1 The principle of superposition

Let us consider an oscillating string fixed at both ends to explain the important principle of periodic motion. A continuous train of sinusoidal waves is traveling back and forth to produce standing waves under an appropriate condition of tension on the string and its length. Figures 1.1–1.3 are snapshots of theses standing waves. Each of the standing waves corresponds to a normal mode of motion. Because the string is fixed at both ends, both ends must be nodes with motion. With the fixed boundary condition, adjacent nodes are one-half wavelength apart and the length of the string may be any integer number of one-half wavelengths. The frequency that gives the longest wavelength is called the first harmonic mode or the fundamental mode, and the others are integral multiples of the fundamental, called the higher harmonics.

**Figure 1.1.** First harmonic (fundamental) mode.

**Figure 1.2.** Second harmonic mode.

**Figure 1.3.** Third harmonic mode.

When a string is initially struck for a short time, the subsequent oscillating string will be described by a linear combination of normal modes. In other words, there are multiple motions of different frequencies on an arbitrarily oscillating string, and the displacement of the arbitrary point of the string is the algebraic sum of the wave

displacements of propagation and reflection. This is called the principle of super-position. While it is a simple statement, it is an essential principle in periodic motion. Let us see how the superposition principle appears in mathematics.

## 1.2 One-dimensional standing wave

The normal modes are the solutions of the linear differential equation for the wave motion with a particular boundary condition

$$\frac{\partial^2 u(x, t)}{\partial t^2} = v^2 \frac{\partial^2 u(x, t)}{\partial x^2} \tag{1.1}$$

where $u(x, t)$ is the displacement, $v = (F/\mu)^{1/2}$ is the wave speed, $F$ is the tension in the string, and $\mu$ is line density of the string. By separating the variables, $u(x, t) = U(x)\Gamma(t)$, the equation for the $x$-coordinate, $U(x)$, becomes the Helmholtz equation whereas the one for time, $\Gamma(t)$, is an equation of harmonic oscillation:

$$\frac{d^2 U(x)}{dx^2} + \lambda U(x) = 0 \text{ and } \frac{d^2 \Gamma(x)}{dt^2} \lambda v^2 \Gamma(t) = 0 \tag{1.2}$$

with the boundary condition of fixed ends, $U(0) = U(L) = 0$.
  (1) Solutions of $U(x)$ are $\sin(n\pi x/L)$, where $\lambda_n = (n\pi/L)^2$ and $n = 1, 2, 3, \ldots$. By the superposition principle, a general solution or an arbitrary wave form with the same boundary condition is given by

$$U(x) = \sum_{n=1}^{\infty} a_n \sin\left(\frac{n\pi}{L} x\right). \tag{1.3}$$

  (2) The solution of $\Gamma(t)$ for a given $\lambda_n$ is a harmonic oscillation and is given by

$$\Gamma_n(t) = C_n \sin\left[\left(\frac{n\pi}{L} v\right) t + \delta_n\right].$$

Therefore,

$$u(x, t) = \sum_{n=1}^{\infty} A_n \sin\left(\frac{n\pi}{L} x\right) \sin\left(\frac{n\pi v}{L} t + \delta_n\right) \tag{1.4}$$

where $A_n = a_n \cdot C_n$ and $\delta$ are determined by a given initial condition.

## 1.3 Fourier series

The previous section describes wave phenomena using linear differential equations where their solutions are given by sinusoidal periodic functions, and an arbitrary wave pattern on the string can be expressed by the superposition of sinusoidal harmonic modes. Likewise, we should be able express an arbitrary periodic function $f(\xi)$, where the variable $\xi$ is a spatial coordinate or time, as a series of sinusoidal functions. This is called the Fourier series of $f(\xi)$. Because the sinusoidal functions

are well-known and easy to apply, Fourier series are very useful for analyzing periodic motions such as wave phenomena.

### 1.3.1 Fourier theorem

In the following discussion, we use the periodicity of time. A periodic function $f(t)$ of period $T$ $(-T/2 < t \leqslant +T/2)$ can be expressed by a Fourier series [1]:

$$f(t) = \frac{a_0}{2} + \sum_{n=1}^{\infty} a_n \cos(n\omega_0 t) + \sum_{n=1}^{\infty} b_n \sin(n\omega_0 t) \qquad (1.5)$$

where $\omega_0 = 2\pi/T$ is the angular frequency of the fundamental mode, and the Fourier coefficients $a_m$ and $b_m$ are given by

$$a_m = \frac{2}{T} \int_{--T/2}^{T/2} f(t)\cos(m\omega_0 t)dt \ \text{ and } \ b_m = \frac{2}{T} \int_{-T/2}^{+T/2} f(t)\sin(m\omega_0 t)dt. \qquad (1.6)$$

The Fourier coefficients are calculated using the orthogonal property of sinusoidal functions.

Using the inner product, $\langle \dots | \dots \rangle$, defined by equation (1.17), we obtain

$$\langle \cos(m\omega_0 t) | \cos(n\omega_0 t) \rangle = \int_{-T/2}^{T/2} \cos(m\omega_0 t)\cos(n\omega_0 t)dt = 0 \ (m \neq n)$$

$$\langle \sin(m\omega_0 t) | \sin(n\omega_0 t) \rangle = \int_{-T/2}^{T/2} \sin(m\omega_0 t)\sin(n\omega_0 t)dt = 0 \ (m \neq n)$$

and

$$\langle \cos(m\omega_0 t) | \sin(n\omega_0 t) \rangle = \int_{-T/2}^{T/2} \cos(m\omega_0 t)\sin(n\omega_0 t)dt = 0 \ (\text{including } m = n).$$

Depending on the property of the original periodic function $f(t)$, a Fourier series may have only sine-terms or cosine-terms. If the function $f(t)$ is an *even* function in the interval $[-T/2, +T/2]$, the sine-terms must be excluded, and the Fourier series has only cosine-terms:

$$f(t) = \frac{a_0}{2} + \sum_{n=1}^{\infty} a_n \cos(n\omega_0 t). \qquad (1.7)$$

Similarly, if the periodic function $f(t)$ is an *odd* function in the interval $[-T/2, +T/2]$, the Fourier series has only sine-terms:

$$f(t) = \sum_{n=1}^{\infty} b_n \sin(n\omega_0 t). \qquad (1.8)$$

We can also obtain a Fourier series in a complex exponential form. By applying Euler's formula, $e^{i\theta} = \cos\theta + i \sin\theta$, the Fourier series (1.5) becomes

$$f(t) = \frac{a_0}{2} + \sum_{n=1}^{\infty} \frac{a_n}{2}[e^{in\omega_0 t} + e^{-in\omega_0 t}] + \sum_{n=1}^{\infty} \frac{b_n}{2i}[e^{in\omega_0 t} - e^{-in\omega_0 t}]$$

$$= \frac{a_0}{2} + \sum_{n=1}^{\infty} \frac{1}{2}(a_n - ib_n)e^{in\omega_0 t} + \sum_{n=1}^{\infty} \frac{1}{2}(a_n + ib_n)e^{-in\omega_0 t} \qquad (1.9)$$

$$= \frac{a_0}{2} + \sum_{n=1}^{\infty} \frac{1}{2}(a_n - ib_n)e^{in\omega_0 t} + \sum_{n=-1}^{-\infty} \frac{1}{2}(a_{-n} + ib_{-n})e^{-in\omega_0 t} = \sum_{n=-\infty}^{\infty} c_n e^{in\omega_0 t}$$

$$\text{where } c_n = \frac{1}{2}(a_n - ib_n) = \frac{2}{T}\int_{-T}^{T} f(t)\{\cos(n\omega_0 t) - i\sin(n\omega_0 t)\}dt = \frac{1}{T}\int_{-L}^{L} f(t)e^{in\omega_0 t}dt \quad (1.10)$$

for $n = 0, 1, 2, \ldots$. For $n < 0$, we can use the same equation (1.10) by defining $a_{-n} = a_n$ and $b_{-n} = b_n$. The complex Fourier coefficient represents the magnitude of the frequency component and the phase.

Instead of a periodic function in time, a Fourier series can be also applied to a periodic function of coordinates. The spatial periodicity can be observed through electromagnetic waves including optical interferences and diffractions, and wave packets of a particle. An arbitrary periodic function $f(x)$ in the interval $[-\pi, +\pi]$

$$f(x) = \frac{a_0}{2} + \sum_{n=1}^{\infty} a_n \cos(nx) + \sum_{n=1}^{\infty} b_n \sin(nx) \qquad (1.11)$$

where the coefficients $\{a_n; n = 0, 1, 2, 3, \ldots\}$ and $\{b_n; n = 1, 2, 3 \ldots\}$ are given by

$$a_n = \frac{1}{\pi}\int_{-\pi}^{\pi} f(x)\cos(nx)dx \text{ and } b_n = \frac{1}{\pi}\int_{-\pi}^{\pi} f(x)\sin(nx)dx. \qquad (1.12)$$

The Fourier series for the spatial interval $[-\pi, +\pi]$ can be changed to $[-L, +L]$ by using another variable $\xi = (L/\pi)x$ or $x = (\pi/L)\xi$

$$f(\xi) = \frac{a_0}{2} + \sum_{n=1}^{\infty} a_n \cos\left(\frac{n\pi}{L}\xi\right) + \sum_{n=1}^{\infty} b_n \sin\left(\frac{n\pi}{L}\xi\right) \qquad (1.13)$$

where

$$a_n = \frac{1}{L}\int_{-L}^{L} f(\xi) \cos\left(\frac{n\pi}{L}\xi\right)d\xi \text{ and } b_n = \frac{1}{L}\int_{-L}^{L} f(\xi) \sin\left(\frac{n\pi}{L}\xi\right)d\xi. \qquad (1.14)$$

If we use complex functions, Fourier series (1.13) becomes

$$f(\xi) = \sum_{n=-\infty}^{\infty} c_k e^{i\pi n\xi/L} \qquad (1.15)$$

$$\text{where } c_n = \frac{1}{2L}\int_{-L}^{L} f(\xi)\exp\left(-i\frac{n\pi}{L}\xi\right)d\xi. \qquad (1.16)$$

**Example 1:** the Fourier series of a square wave. The first example of the Fourier series is a square wave train:

$$f(x) = -1 \text{ if } -\pi < x < 0; \text{ and } +1 \text{ if } 0 < x < \pi.$$

The graph of this square wave train for $x > 0$ is shown in figure 1.4. The Fourier series of the square wave train is given by:

$$f(x) = \frac{4}{\pi} \sum_{n=1}^{\infty} \frac{1}{(2n-1)} \sin[(2n-1)x] = \frac{4}{\pi}\left(\sin x + \frac{1}{3}\sin 3x + \frac{1}{5}\sin 5x + \frac{1}{7}\sin 7x + ....\right).$$

Figure 1.5 is the actual Fourier series of the first 10 terms—except the multiplication factor $\pi/4$—by iterative summation using EXCEL®. Its macro source program using the built-in Visual Basic for Applications (VBA) is listed in the appendix, A13.1. If readers are not familiar with EXCEL VBA macro, refer to A12.3 in the appendix.

If we calculate the sum of 100 terms, the Fourier series gets much closer to the square wave train with noticeable oscillations at the rising and falling edges (figure 1.6).

**Remark:** Gibbs phenomenon. The fine oscillations do not disappear, especially at the edges, even if the Fourier series takes many more terms. This overshoot behavior occurs at a jump discontinuity, and it is called the Gibbs phenomenon [2]. The size of the overshoot is tuned out to be proportional to the magnitude of the discontinuity. For the square wave train, the maximum peak value of the partial sum approaches approximately $\frac{2}{\pi}\int_0^\pi \frac{\sin \xi}{\xi}d\xi \approx 1.1789$.

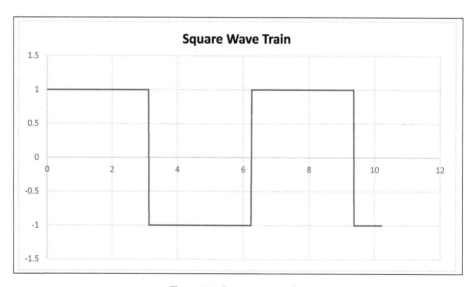

**Figure 1.4.** Square wave train.

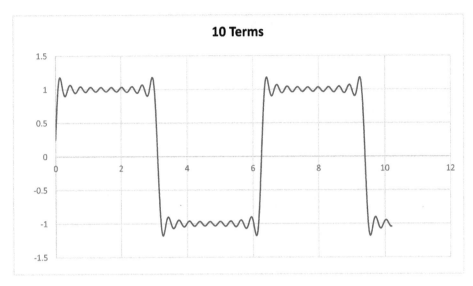

**Figure 1.5.** Fourier series of square wave train (10 terms).

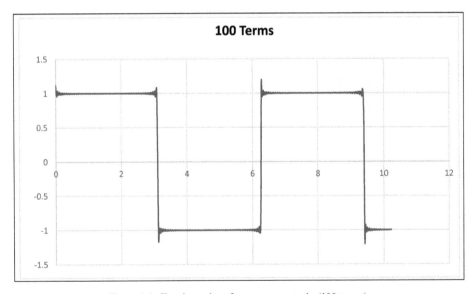

**Figure 1.6.** Fourier series of square wave train (100 terms).

Its calculation offers a good mathematical exercise and readers should try it. Refer to appendix A1 for a detailed calculation.

**Example 2:** the Fourier series of a sawtooth wave. The sawtooth wave is a repetition of the function $f(t) = x$ for $-\pi < x < +\pi$, and the period is $2\pi$. Figure 1.7 shows this signal for $x \geqslant 0$. The Fourier series of the above sawtooth wave is

**Figure 1.7.** Sawtooth wave.

**Figure 1.8.** Fourier transform of sawtooth wave (10 terms).

$$f(t) = \sum_{n=1}^{\infty} \frac{2(-1)^{n+1}}{n} \sin(nx) = 2\left[ \sin x - \frac{1}{2} \sin 2x + \frac{1}{3} \sin 3x - \frac{1}{4} \sin 4x + -\cdots \right].$$

The Fourier series up to 10 terms and 100 terms are shown in figures 1.8 and 1.9, respectively. The Gibbs phenomenon is also noticeable in this Fourier series.

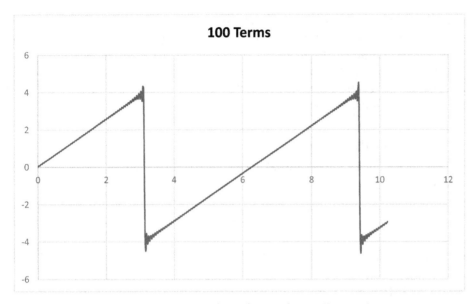

**Figure 1.9.** Fourier transform of sawtooth wave (100 terms).

## 1.4 Orthonormal basis

The concept of orthonormal basis is the foundation of the superposition principle and the theory of the Fourier series. Let us consider an $n$-dimensional Euclidian vector space for an intuitive discussion of the orthonormal basis. In this space, an arbitrary vector can be expressed as a linear combination of unit vectors:

$$\vec{v} = v_1\vec{e_1} + v_2\vec{e_2} + \cdots + v_n\vec{e_n} \tag{1.17}$$

where $\{v_j;\ j\ =\ 1,\ 2,\ ....,\ n\}$ are the components of the vector $\vec{v}$, and $\{\vec{e_j}; j = 1, 2, \ldots, n\}$ form a set of unit vectors associated with the given Cartesian coordinate frame. The component $v_j$ is given by the inner product of the vector and the unit vector

$$v_i = \langle\vec{v}|\vec{e}\rangle = \sum_{j=1}^{n} v_j\langle\vec{e_j}|\vec{e_i}\rangle \tag{1.18}$$

because the unit vectors are orthonormal: $\langle\vec{e_j}|\ \vec{e_i}\rangle\ =\ \delta_{ij}$ where $\delta_{ij}$ is the Kronecker's delta, i.e., $\delta_{ij} = 1$ if $i = j$ and 0 otherwise. Furthermore, there are no other additional unit vectors required to express an arbitrary vector in the $n$-dimensional space. Thus, the unit vectors $\{\vec{e_i}\ ;\ i\ =\ 1, 2, \ldots, n\}$ form a complete orthonormal basis of the vector space.

As discussed in section 1.2, the standing wave equation has sinusoidal solutions as the normal modes in the given Cartesian coordinates. Because sinusoidal functions are orthogonal, for the string oscillation, the normal modes can be regarded as the

unit vectors and an arbitrary string wave can be expressed in the form of the linear combination (superposition) of the normal modes, constituting the Fourier series. This is what equation (1.3) implies.

A time-dependence of temperature or particle diffusion in a rod computed from a heat or diffusion equation are other examples where we can apply Fourier series. Let us discuss an example below.

## 1.5 Heat and diffusion equations

Consider the following heat transfer or particle diffusion problem given by the heat equation:

$$\frac{\partial u(x,\ t)}{\partial t} = \kappa \frac{\partial^2 u(x,\ t)}{\partial x^2} \tag{1.19}$$

where $u(x,\ t)$ is the temperature distribution and $\kappa$ is the thermal diffusivity. For particle diffusion, $u(x,\ t)$ is the particle density distribution and $\kappa$ is the diffusion constant. The spatial part, $X(x)$, of the above equations also becomes the Helmholtz equation (1.2).

Let us find the solution of the heat equation with the following initial and the boundary conditions:

    (i) initial condition: $u(x,\ 0) = f(x)$ for $0 < x < L$, and $u(0,0) = u(L,\ 0) = 0$; and
    (ii) boundary condition: $u(0,\ t) = u(L,\ t) = 0$.

By setting $u(x,\ t) = \Gamma(t)X(x)$ to separate the variables $x$ and $t$, we obtain $X'(x) + \lambda X(x) = 0$ and $\Gamma'(t) + \lambda \kappa \Gamma(t) = 0$ where $\lambda$ is the separation constant.

    (1) $d^2 X(x)/dx^2 + \lambda X(x) = 0$ : using the boundary condition, $U(0) = U(L) = 0$, we obtain
        $X_n(x) = c_1 \sin(\frac{n\pi}{L}x)$ where $c_1$ is a constant, and $\lambda_n = \frac{n\pi}{L}$ where $n = 1, 2, 3,$
        …
    (2) $d\Gamma(t)/dt + \lambda \kappa \Gamma(t) = 0$: the solution is $\Gamma_n(t) = c_2 \exp[-D(\frac{n\pi}{L})^2 t]$ for each $\lambda_n$.

Thus, the temperature distribution, $u(x,\ t)$, is given by

$$u(x,\ t) = \sum_n \Gamma_n(t) X_n(x) = \sum_n A_n \exp\left[-\kappa\left(\frac{n\pi}{L}\right)^2 t\right] \sin\left(\frac{n\pi}{L}x\right) \tag{1.20}$$

where $A_n = c_1 c_2$.

Now, we can include the initial condition, $u(x,\ 0) = f(x)$ by using the Fourier series

$$u(x,\ 0) = f(x) = \sum_n A_n \sin\left(\frac{n\pi}{L}x\right) \tag{1.21}$$

where the coefficient $A_n$ is given by $A_n = \frac{2}{L}\int_0^L f(x)\sin(\frac{n\pi}{L})dx$. Therefore, the complete solution becomes

**Figure 1.10.** Temperature distribution in a rod.

$$u(x,\ t) = \frac{2}{L}\sum_{n}\left[\int_{0}^{L}f(x)\sin\left(\frac{n\pi}{L}\right)dx\right]\exp\left[-\kappa\left(\frac{n\pi}{L}\right)^{2}t\right]\sin\left(\frac{n\pi}{L}x\right). \qquad (1.22)$$

Figure 1.10 shows the time-dependence of the temperature distribution in a rod of length $L = \pi$, where both ends ($x = 0$ and $\pi$) are in contact with heat reservoirs of zero temperature. The rod is at a uniform temperature $f(x) = 100$ for $0<x<L$ at $t = 0$.

## 1.6 Two-dimensional standing wave and two-dimensional Fourier series

**Two-dimensional standing wave:** this is similar to the standing wave on a string discussed in section 1.2, the standing wave on a membrane is an example of a two-dimensional Fourier series. Let us discuss how to obtain two-dimensional standing waves. The wave equation on a membrane is

$$\frac{\partial^{2}u(x,\ y,\ t)}{\partial t^{2}} = v^{2}\left[\frac{\partial^{2}u(x,\ y,\ t)}{\partial x^{2}} + \frac{\partial^{2}u(x,\ y,\ t)}{\partial y^{2}}\right]. \qquad (1.23)$$

The solution of equation (1.23) can be obtained by the variable separation. Let $u(x, y, t) = X(x)Y(y)\Gamma(t)$, then equation (1.23) becomes

$$\frac{X''(x)}{X(x)} + \frac{Y''(y)}{Y(y)} = \frac{1}{v^{2}}\frac{\Gamma''(t)}{\Gamma(t)} \qquad (1.24)$$

and we have

$$\frac{X''(x)}{X(x)} = -k_x^2 \text{ and } \frac{Y''(y)}{Y(y)} = -k_y^2. \tag{1.25}$$

Assume the following initial and boundary conditions for a rectangular membrane:

(i) initial condition: $u(x, y, 0) = f_1(x, y)$ and $(\partial u/\partial t)_{t=0} = f_2(x, y)$; and
(ii) boundary condition: $u(0, y, t) = u(a, y, t) = 0$ and $u(x, 0, t) = u(x, b, t) = 0$.

The solutions of equation (1.24) including the boundary condition are

$$X_n(x) = A_n \sin\left(\frac{m\pi}{a}x\right) \text{ where } k_x = \frac{m\pi}{a} \text{ and } m = 1, 2, 3, \dots ; \tag{1.26}$$

$$Y_n(y) = B_n \sin\left(\frac{n\pi}{b}y\right) \text{ where } k_x = \frac{n\pi}{a} \text{ and } n = 1, 2, 3, \dots ;$$

and the time part is a harmonic oscillation given by

$$\frac{d^2\Gamma(t)}{dt^2} + \omega^2\Gamma(t) = 0 \text{ where } \omega^2 = v^2\left(k_x^2 + k_y^2\right). \tag{1.27}$$

Because possible $k_x$ and $k_y$ values are discrete, $\omega$ is also discrete. We obtain

$$\omega_{m, n}^2 = v^2\pi^2\left[\left(\frac{m}{a}\right)^2 + \left(\frac{n}{b}\right)^2\right], \text{ and thus } \Gamma_{m, n}(t) = C_{m, n} \sin(\omega_{m, n}t + \varepsilon_{m, n}). \tag{1.28}$$

The general solution is now given by

$$u_{m, n}(x, y, t) = \sum_{m=1}^{\infty}\sum_{n=1}^{\infty} X_m(x) Y_n(y)\Gamma_{m, n}(t)$$

$$= \sum_{m=1}^{\infty}\sum_{n=1}^{\infty} D_{m, n} \sin\left(\frac{m\pi}{a}x\right)\sin\left(\frac{n\pi}{b}y\right)\sin(\omega_{m, n}t + \varepsilon_{m, n}) \tag{1.29}$$

where $\{D_{m,n}\} = \{A_m B_n C_{m,n}; m = 1,2,3, \dots, \text{ and } n = 1,2,3,\dots\}$ and $\varepsilon_{m,n}$ are determined by the initial condition. The spatial part is a two-dimensional (sine) Fourier series. Figure 1.11 shows the spatial part of $u_{3,2}$ with $a = b = 3.15$. Refer to appendix A13.3 for how to draw this 3D chart using EXCEL.

**Note:** a popular demonstration of standing waves on a Chladni plate has the boundary conditions of free ends and it is a forced oscillation because the plates are driven at the center vertically. Therefore, the resonant modes of a Chladni plate are different from equation (1.29). Refer to [3] for more information on Chladni plates.

**Two-dimensional Fourier series:** equation (1.29) is an example of a two-dimensional Fourier series. The two-dimensional Fourier series of an arbitrary periodic function $f(x,y)$ can be formulated in the following way. First, construct a Fourier series with respect to $x$ whence coefficients are function of $y$. Second, construct

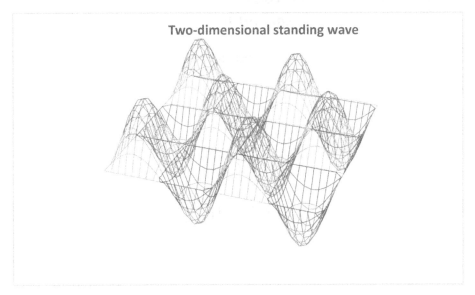

**Figure 1.11.** Two-dimensional standing wave on a membrane.

another Fourier series with respect to $y$ from the coefficients. For example, from the one-dimensional complex Fourier series (1.13), we can formulate two-dimensional complex Fourier series:

$$f(x, y) = \sum_{k=-\infty}^{\infty} \sum_{\ell=-\infty}^{\infty} c(k, \ell) \exp\left[i\pi\left(\frac{kx}{L_x} + \frac{\ell y}{L_y}\right)\right] \tag{1.30}$$

where $c(k, \ell) = \dfrac{1}{4L_xL_y} \int_{-L_x}^{L_x} dx \int_{-L_y}^{L_y} dy \exp\left[-i\pi\left(\dfrac{kx}{L_x} + \dfrac{\ell y}{L_y}\right)\right]$, $k, \ell = 0, \pm 1, \pm 2, \ldots$

Because we do not use the two-dimensional Fourier series, refer to [4] for a general discussion of multi-dimensional Fourier series as necessary.

## References

[1] Stewart J 2016 *Fourier Series—Stewart Calculus* (Boston, MA: Cengage Learning)
[2] Gibbs Phenomena, *MIT Open Course Ware* (https://ocw.mit.edu/courses/mathematics/18-03sc-differential-equations-fall-2011/unit-iii-fourier-series-and-laplace-transform/operations-on-fourier-series/MIT18_03SCF11_s22_7text.pdf)
[3] Georgi H *11.3: Chladni Plates—Physics LibreTexts*
[4] Pivato M 2012 *Linear Partial Differential Equations and Fourier Theory Cambridge Core* (Cambridge: Cambridge University Press) (*Multidimensional Fourier Series (Chapter 9)— Linear Partial Differential Equations and Fourier Theory* (cambridge.org))

IOP Publishing

# Fourier Transform and Its Applications Using Microsoft EXCEL® (Second Edition)

Shinil Cho

# Chapter 2

# The Fourier transform

In this chapter, we define the Fourier transform from the Fourier series, and list the basic properties of the Fourier transform. To acquire the measurement data, we usually sample signals at periodic intervals, and the acquired data is a set of discrete values. Discrete Fourier transforms computed by computers is based on a special algorithm, called the fast Fourier transform (FFT). We describe the essential parts of the FFT. This chapter also has an introduction of linear response theory where the Fourier transform plays an important role. In chapter 4, we will discuss how the theory is applied in various physics subjects.

## 2.1 From the Fourier series to the Fourier transform

### 2.1.1 The concept of the Fourier transform

Because the Fourier series is applied to periodic functions, when we attempt to apply the theory to a non-periodic signal, we need to modify the theory. We can extend the Fourier series to a non-periodic function by making the period of the non-periodic function infinite. This mathematical procedure is called the Fourier transform [1].

The complex Fourier series of a periodic function $f(t)$ of period $-T/2 \leqslant x \leqslant + T/2$ is given by equation (1.9). Including the explicit form of the coefficients $c_n$, we can write

$$f(t) = \sum_{n=-\infty}^{n=+\infty} \left[ \frac{1}{T} \int_{-T/2}^{+T/2} f(\tau)e^{-in\omega_0\tau}d\tau \right] e^{in\omega_0 t} = \sum_{n=-\infty}^{n=+\infty} \left[ \frac{\omega_0}{2\pi} \int_{-T/2}^{+T/2} f(\tau)e^{-in\omega_0\tau}d\tau \right] e^{in\omega_0 t} \quad (2.1)$$

where $\omega_0 = 2\pi/T$ is the angular frequency. It becomes infinitesimally small in the limit of $T \to \infty$. Thus, in this limit, $\omega_0$ can be replaced with $d\omega$, and the integer multiplication of the angular frequency $\omega_0$—given by $\omega = n\omega_0$ where $n$ is the integer —can be regarded as a continuous variable $\omega$. Then, in the limit of $T \to \infty$, the Fourier series of $f(t)$ becomes

doi:10.1088/978-0-7503-6044-9ch2

$$f(t) = \sum_{n=-\infty}^{n=+\infty} \left[ \frac{\omega_0}{2\pi} \int_{-\infty}^{+\infty} f(\tau)e^{-in\omega_0\tau}d\tau \right] e^{in\omega_0 t} \rightarrow \int_{-\infty}^{+\infty} \frac{d\omega}{2\pi} \left[ \int_{-\infty}^{+\infty} f(\tau)e^{-i\omega\tau}d\tau \right] e^{i\omega t}. \quad (2.2)$$

We call the integrant of equation (2.2) the Fourier transform of the function $f(t)$:

$$F(\omega) = \int_{-\infty}^{+\infty} f(t)e^{-i\omega t}dt \quad (2.3)$$

and we call the following equation (2.4) the *inverse* Fourier transform of the function $F(\omega)$:

$$f(t) = \frac{1}{2\pi} \int_{-\infty}^{+\infty} F(\omega)e^{i\omega t}d\omega. \quad (2.4)$$

For the Fourier transform $F(\omega)$ to exist, $F(\omega)$ should be finite everywhere, and thus

$$|F(\omega)| = \left| \int_{-\infty}^{+\infty} f(t)e^{-i\omega t}dt \right| \leqslant \int_{-\infty}^{+\infty} |f(t)e^{-i\omega t}|dt < \infty. \quad (2.5)$$

This finite condition is satisfied if $f(t) \rightarrow 0$ as $t \rightarrow \pm\infty$. With this condition, a non-periodic function, whose asymptotic values are zero, has the Fourier transform. The Fourier transform defined by equations (2.3) and (2.4) connect two different descriptions of a physics system: the time-based and the frequency-based. By the same token, a system described in terms of the spatial coordinates forms a pair of Fourier transform variables with wave number (or momentum in quantum mechanics) $k$:

$$F(k) = \int_{-\infty}^{+\infty} f(x)e^{-ikx}dx \quad (2.6)$$

where

$$f(x) = \frac{1}{2\pi} \int_{-\infty}^{+\infty} F(k)e^{ikx}dk. \quad (2.7)$$

Using the vector notation, for $n$-dimensional FT can be written as

$$F(\vec{k}) = \int_{-\infty}^{\infty} f(\vec{x})e^{-i\vec{k}\cdot\vec{x}}d\vec{x} \text{ and } f(x) = \frac{1}{(2\pi)^n} \int_{-\infty}^{+\infty} F(\vec{k})e^{i\vec{k}\cdot\vec{x}}d\vec{k}. \quad (2.8)$$

**Remark:** in some cases, unlike equations (2.3) and (2.4), the Fourier transform with respect to time may have swapped signs of exponential functions

$$F(\omega) = \int_{-\infty}^{+\infty} f(t)e^{i\omega t}dt \Leftrightarrow f(t) = \frac{1}{2\pi} \int_{-\infty}^{+\infty} F(\omega)e^{-i\omega t}d\omega. \quad (2.9)$$

This is possible if the exponential Fourier series is defined as $f(t) = \sum_{n=-\infty}^{\infty} c_n e^{-in\omega_0 t}$

unlike equation (1.9). This definition would be better for wave phenomena. For example, an arbitrary one-dimensional propagating wave to the positive direction is

given by $f(x-vt)$ where $v$ is the wave speed. If we need to compute Fourier transforms with respect to the coordinate $x$ and time $t$ simultaneously, the definition of equation (2.9) is more natural

$$F(k, \omega) = \int \int f(x - vt)e^{-i(kx-\omega t)}dxdt \iff f(x - vt) = \frac{1}{2\pi} \int \int F(k, \omega)e^{i(kx-\omega t)}dkd\omega. \quad (2.10)$$

### 2.1.2 Mathematical properties of a Fourier transform

There are several important mathematical properties of a Fourier transform as listed below.

1. **Linearity**: suppose a function $f(t)$ is a linear combination of two functions, i.e., $f(t) = af_1(t) + bf_2(t)$ where $a$ and $b$ are constants. The Fourier transform of the function $f(t)$ is given by

$$F(\omega) = \int_{-\infty}^{+\infty} f(\tau)e^{-ikx}d\tau = \int_{-\infty}^{+\infty} (af_1(\tau) + bf_2(\tau))e^{-ikx}d\tau = aF_1(\omega) + bF_2(\omega) \quad (2.11)$$

where $F_1(\omega) = \int_{-\infty}^{+\infty} f_1(\tau)e^{-i\omega\tau}d\tau$ and $F_2(\omega) = \int_{-\infty}^{+\infty} f_2(\tau)e^{-i\omega\tau}d\tau$.

Therefore, a Fourier transform is also *linear*.

2. **Time invariance**: consider the time shift by $t_0$ of a Fourier transform.

$$F'(\omega) = \int_{-\infty}^{+\infty} f(t - t_0)e^{-i\omega t}dt = \int_{-\infty}^{+\infty} f(\tau)e^{-i\omega(\tau+t_0)}d\tau = e^{-i\omega t_0} \int_{-\infty}^{+\infty} f(\tau)e^{-i\omega\tau}d\tau \quad (2.12)$$

where $F(\omega) = \int_{-\infty}^{+\infty} f(t)e^{-i\omega t}dt$.

Thus, $F'(\omega) = e^{-i\omega t_0}F(\omega)$ and $|F'(\omega)| = |F(\omega)|$. Therefore, we say the Fourier spectrum is *time invariant*. By the same token, for the coordinate shift by $x_0$ of a Fourier transform

$$F'(k) = \int_{-\infty}^{\infty} f(x - x_0)e^{-ikx}dx = \int_{-\infty}^{\infty} f(\xi)e^{-ik(\xi+x_0)}d\xi = e^{-ikx_0} \int_{-\infty}^{\infty} f(\xi)e^{-ik\xi}d\xi = e^{-ikx_0}F(k)$$

where

$$F(k) = \int_{-\infty}^{\infty} f(\xi)e^{-ik\xi}d\xi. \text{ Thus, } |F'(k)| = |F(k)|.$$

Therefore, the Fourier transform is space invariant.

3. **Convolution**: a Fourier transform of the convoluted integral. The convoluted integral is defined as $f(t) \times g(t) \equiv \int_{-\infty}^{+\infty} f(\tau)g(t - \tau)d\tau$. Its Fourier transform is given by

$$\int_{-\infty}^{+\infty} \left[ \int_{-\infty}^{+\infty} f(\tau)g(t - \tau)d\tau \right]e^{-i\omega t}dt = \int_{-\infty}^{+\infty} \left[ \int_{-\infty}^{+\infty} f(\tau)g(t - \tau)d\tau \right]e^{-i\omega(t-\tau)-i\omega\tau}dt$$

$$= \int_{-\infty}^{+\infty} f(\tau)e^{-i\omega\tau}d\tau \int_{-\infty}^{+\infty} g(t - \tau)e^{-i\omega(t-\tau)}dt = F(\omega)G(\omega). \quad (2.13)$$

We also obtain

$$\int_{-\infty}^{+\infty} f(\tau)g(t-\tau)d\tau = \frac{1}{2\pi} \int_{-\infty}^{+\infty} f(\tau)\left[\int_{-\infty}^{+\infty} G(\omega)e^{i\omega(t-\tau)}d\omega\right]d\tau$$

$$= \frac{1}{2\pi} \int_{-\infty}^{+\infty} f(\tau)e^{-i\omega\tau}d\tau \int_{-\infty}^{+\infty} G(\omega)e^{i\omega t}d\omega = \int_{-\infty}^{+\infty} F(\omega)G(\omega)e^{i\omega t}d\omega.$$
(2.14)

In particular, taking $t = 0$ in the above equation, we obtain

$$\int_{-\infty}^{+\infty} f(\tau)g(-\tau)d\tau = \int_{-\infty}^{+\infty} F(\omega)G(\omega)e^{i\omega t}d\omega.$$
(2.15)

Equations (2.13) to (2.15) represent the convolution theorem where a Fourier transform of the convolution integral is a product of two separate Fourier transforms.

4. **Parseval's equality:**

$$\int_{-\infty}^{+\infty} |f(t)|^2 dt = \int_{-\infty}^{+\infty} dt \frac{1}{2\pi} \int_{-\infty}^{+\infty} F^*(\omega)\, e^{-i\omega t}d\omega \frac{1}{2\pi} \int_{-\infty}^{+\infty} F(\omega')e^{i\omega' t}d\omega'$$

$$= \frac{1}{2\pi} \int_{-\infty}^{+\infty} F^*(\omega)d\omega \int_{-\infty}^{+\infty} F(\omega')\delta(\omega' - \omega)d\omega' = \frac{1}{2\pi} \int_{-\infty}^{+\infty} F^*(\omega)F(\omega)d\omega$$
(2.16)

where $\frac{1}{2\pi} \int e^{i(\omega'-\omega)t}dt = \delta(\omega' - \omega)$ is the Dirac delta function described as a mathematics property (6) in this list. Therefore, we obtain Parseval's theorem:

$$\int_{-\infty}^{+\infty} |f(t)|^2 dt = \frac{1}{2\pi} \int_{-\infty}^{+\infty} |F(\omega)^2|d\omega.$$
(2.17)

When $f(t)$ is a signal of a physical system, $|f(t)|^2$ is proportional to the energy of the signal, and therefore Parseval's theorem implies that the total energy in the time domain is equal to that in the frequency domain.

5. **Hermitian:** $F^*(\omega) = F(\omega)$. If $f(t)$ is a real-value function, i.e., $f^*(t) = f(t)$,

$$f(x) = \frac{1}{2\pi} \int_{-\infty}^{+\infty} F(k)e^{-ikx}dk \text{ and}$$

$$f^*(x) = \frac{1}{2\pi} \int_{-\infty}^{+\infty} F^*(k)e^{-ikx}dk = \frac{1}{2\pi} \int_{-\infty}^{+\infty} F^*(-k)e^{ikx}dk.$$

Therefore, the Fourier transform of a real-time function is *Hermitian*: $F^* (\omega) = F(\omega)$.

6. **The Fourier transform of the Dirac delta function:** Combining the Fourier transform and its inverse transform, we obtain

$$f(x) = \frac{1}{2\pi} \int_{-\infty}^{+\infty} F(k)e^{ikx}dk = \frac{1}{2\pi} \int_{-\infty}^{+\infty} dk e^{ikx}\left(\int_{-\infty}^{+\infty} dx' f(x')e^{-ikx'}\right)$$

$$= \int_{-\infty}^{+\infty} dx' f(x')\left(\frac{1}{2\pi} \int_{-\infty}^{+\infty} dk e^{ik(x-x')}\right).$$

Thus, the delta function can be given by

$$\delta(x - x') = \frac{1}{2\pi} \int_{-\infty}^{+\infty} dk e^{ik(x-x')}. \tag{2.18}$$

Comparing this equation with the definition of a Fourier transform, the Fourier transform of the function $D(k)$ is the delta function.

$$\delta(x) = \int_{-\infty}^{+\infty} D(k)e^{ikx}dk \text{ where } D(k) = \frac{1}{2\pi}. \tag{2.19}$$

7. **The Fourier transform of a step function:** consider a step function: $u(t) = 0$ if $x < 0$; $0.5$ if $x = 0$; and $1$ if $x > 0$.
   Its Fourier transform is computed to

$$U(\omega) = \int_{0}^{+\infty} e^{i\omega t}dt = \pi\delta(\omega) - \frac{i}{\omega}. \tag{2.20}$$

Appendices A2 and A3 show how to calculate the integral.

8. **Differentiation:** if $F(\omega) = \int_{-\infty}^{+\infty} f(t)e^{-i\omega t}dt$, then

$$\int_{-\infty}^{+\infty} \frac{df(t)}{dt}e^{-i\omega t}dt = [f(t)e^{-i\omega t}]_{-\infty}^{+\infty} + i\omega \int_{-\infty}^{+\infty} f(t)e^{-i\omega t}dt = i\omega F(\omega) \tag{2.21}$$

where $f(t) \to 0$ as $t \to \pm\infty$.

9. **Integration:** using the step function $u(t)$ defined in the mathematics property (7),

$$\int_{-\infty}^{t} f(\tau)d\tau = \int_{-\infty}^{+\infty} f(\tau)u(t - \tau)d\tau.$$

This is the convolution integral discussed in property (7), and the Fourier transform of the above integral can be given by

$$\int_{-\infty}^{+\infty} \left( \int_{-\infty}^{t} f(\tau)d\tau \right) e^{-i\omega t}dt = F(\omega)\left[ \pi\delta(\omega) - \frac{i}{\omega} \right] = \pi F(\omega)\delta(\omega) - \frac{i}{\omega}F(\omega). \tag{2.22}$$

## 2.2 Practical computational issues of the Fourier transform

While a signal $f(t)$ is generally a continuous function of time, sampling the signal is carried out only at discrete time steps in measurement. This gives rise to the following issues when calculating Fourier transforms:

(1) slow sampling rate—the Nyquist theory states that the rate of acquiring the signal (sampling frequency) must be faster than the maximum frequency of the signal by a factor of two or more;
(2) aliasing—the appearance of ghost signals as a consequence of slow sampling rate; and

(3) finite size sample—a data set of finite size may generate non-periodic components, generating unwanted side-bands in the frequency domain.

Because these issues need to be examined carefully for more accurate Fourier transforms, what follows is a discussion of how to suppress them.

### 2.2.1 Sampling rate slower than signal frequency

There is a maximum frequency that laboratory instruments can measure because their master clocks cannot follow input signals exceeding the maximum frequency of the master clock frequency. The Nyquist-Shannon sampling theorem states that, in order to measure a signal of frequency $f_c$, the equipment that measures the signal must sample the signal at the frequency $2f_c$ (the Nyquist frequency) or higher [2]. Figure 2.1 illustrates that the signal frequency, as shown in red, is exactly twice the sampling frequency, as shown in blue. In this case, since we only measure the signal peaks at every other interval, the signal frequency measured is the same as the sampling rate. In other words, we cannot measure the signal frequency correctly unless the sampling frequency is at least twice the signal frequency. For this reason, the master oscillator frequency of an oscilloscope is usually 5 to 10 times higher than the specified maximum measurable frequency.

### 2.2.2 Aliasing (folding noise)

If there is a signal component of a higher frequency than the sampling frequency, what will actually happen? The signal component observed will exhibit a different frequency. For example, if we sample a sine wave with a frequency of 5 Hz or a period of 0.20 s with the sampling rate of 0.25 s, the period of the sampled signal will be their least common multiple, 1.0 s. Figure 2.2 shows how a signal of $y = \sin(10\pi t)$

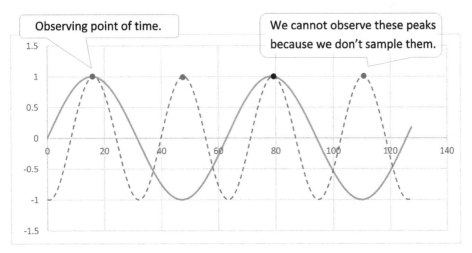

**Figure 2.1.** Nyquist–Shannon sampling theorem. Remark: **Blue**—sampling frequency, and **Red**—signal frequency to be observed. ●: measurable points; ●: non-measurable points.

**Figure 2.2.** Aliasing of sine-signal of period of 0.20 sampled at rate of 0.25.

**Figure 2.3.** A signal on an oscilloscope screen with a slower sweep time.

is measured with the sampling time interval of 0.25 s. The sampled data indicates the false period of 1 s!

This false signal is called aliasing, and it is similar to what we will see of a signal on an oscilloscope when the sweep time of the oscilloscope is slower that the signal period. It is also called the Moiré pattern or signal overlapping. Figure 2.3 shows a digital oscilloscope screen analogous to aliasing.

### 2.2.3 A finite size sample

A finite length of measurement time truncates the signal length. It may induce a non-periodic component within the measurement time interval. As a result, artificial ghosts would appear around the spectral peak. In a Fourier transform, the truncation is mathematically equivalent to a multiplication of the signal with a

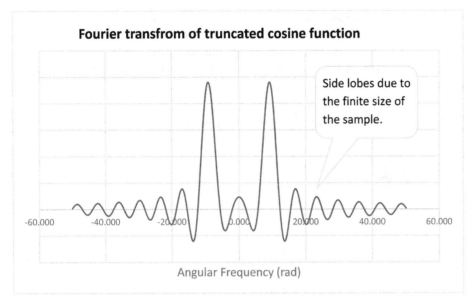

**Figure 2.4.** Fourier transform of the truncated cosine function.

rectangular window. That is, the signal to be Fourier transformed is a product of $f(t)$ and $w(t)$ where $w(t)$ is a rectangular window. To illustrate the effect of a finite size sample, consider the Fourier transform of a $\cos(\omega_0 t)$:

$$C(\omega) = \int_{-\infty}^{+\infty} \cos(\omega_0 t)e^{i\omega t}dt = \frac{1}{2}[\delta(\omega + \omega_0) + \delta(\omega - \omega_0)]. \quad (2.23)$$

This is the sum of two Dirac delta functions at $\omega = \pm\omega_0$. Calculating the Fourier transform of the truncated cosine function, we obtain

$$F(\omega) = \int_{-T}^{+T} \cos(\omega_0 t)e^{-i\omega t}dt = T\left[\frac{\sin[(\omega - \omega_0)T]}{(\omega - \omega_0)T} + \frac{\sin[(\omega + \omega_0)T]}{(\omega + \omega_0)T}\right]. \quad (2.24)$$

Figure 2.4 shows the Fourier transform (where $T = 1$ and $\omega_0 = 9.0$). As seen in this figure, there are two peaks at $\omega = \pm\omega_0$ accompanied by many side lobes. Due to the side lobes, higher harmonic ghosts appear in the spectrum.

Because we sample signals for finite lengths of time, they are always accompanied by side lobes. Several windowing functions have been proposed to suppress this unwanted distortion. In chapter 3, we will demonstrate how these windowing functions affect energy spectra.

## 2.3 Discrete Fourier transform and fast Fourier transform

A set of discrete signals $\{x(t)\}$ sampled from a continuous signal $x_s(t)$ is given by

$$x_s(t) = \sum_{n=0}^{N} x(t)\delta(t - n\Delta t) \quad (2.25)$$

where $\delta$ is the delta function, $\Delta t$ is the sampling period, and $N$ is the total number of sampling points. The time duration of the sampled signal is thus $T = N\Delta t$. For simplicity, we denote the set of sampled signals as $\{x(n); n = 0, 1, 2, ..., N-1\}$.

The discrete Fourier transform (DFT) is defined with the discrete signal (2.25) to become

$$X(\omega) = \int_{-\infty}^{\infty} x_s(t)e^{-i\omega t}dt = \sum_{n=0}^{N-1}\int_{-\infty}^{\infty} x(t)\delta(t - n\Delta t)e^{-i\omega t}dt = \sum_{n=0}^{N-1} x(n\Delta t)e^{-i\omega n\Delta t}$$

$$X(k\omega_0) = \sum_{n=0}^{N-1} x(n\Delta t)e^{-ik\cdot 2\pi n/N}$$

or simply

$$X(k) = \sum_{n=0}^{N-1} x(n)W_N^{kn} \tag{2.26}$$

where $\omega = k\cdot\omega_0$, $\omega_0 = 2\pi/T = 2\pi/N\Delta t$, $W_N = e^{-i\frac{2\pi}{N}}$, and $k = 0,1, 2, ..., N-1$. The inverse Fourier transform is given by

$$x(n) = \frac{1}{N}\sum_{k=0}^{N-1} X(k)e^{i2\pi\frac{kn}{N}} = \frac{1}{N}\sum_{k=0}^{N-1} X(k)W_N^{-kn}. \tag{2.27}$$

**Remark:** if the total number of data sampled at time interval $T$ is $N$, the total measurement time interval is $NT$, and thus the frequency resolution becomes $\Delta f = 1/NT$.

### 2.3.1 $N$-point FT from two ($N/2$)-point FTs

The DFT is suitable for computation using a computer. The number of terms to compute for a $N$-point FT is of the order of $N^2$. When the computational speed of computers was slow, the Fourier transform was a very time-consuming task. Cooley and Tukey developed an efficient method for performing the Fourier transform, known as the fast Fourier transform (FFT) [3, 4]. The new algorithm efficiently uses already-calculated terms for the next terms to be calculated using the periodicity of terms $W_N^{kn} = e^{-i2\pi kn/N}$ and reduces the number of calculation terms to the order of $N\cdot\log N$ from $N^2$.

We can apply the idea of the FFT to construct the $N$-point FT from two $N/2$-point FTs. First, we rewrite the $N$-point FT as the sum of the even and odd points FTs.

$$X(k) = \sum_{n=0}^{N-1} x(n)W_N^{kn} = \sum_{m=0}^{(N/2)-1} x(2m)W_N^{2mk} + \sum_{m=0}^{(N/2)-1} x(2m + 1)W_N^{(2m+1)k} \tag{2.28}$$

where $W_N^{kn} = e^{-i2\pi kn/N}$. Notice that because of the periodicity of exponential function, we find

$$W_N^{2mk} = e^{-i\frac{2\pi}{N}2mk} = e^{-i\frac{2\pi}{(N/2)}mk} = W_{\frac{N}{2}}^{mk} \text{ and } W_N^{(2m+1)k} = W_N^k W_N^{2mk} = W_N^k W_{\frac{N}{2}}^{mk}.$$

Thus, equation (2.28) becomes

$$X(k) = \sum_{m=0}^{(N/2)-1} x(2m)W_N^{2mk} + \sum_{m=0}^{(N/2)-1} x(2m+1)W_N^{(2m+1)k}$$

$$= \sum_{m=0}^{(N/2)-1} x(2m)W_{\frac{N}{2}}^{mk} + W_N^k \sum_{m=0}^{(N/2)-1} x(2m+1)W_{\frac{N}{2}}^{mk}. \tag{2.29}$$

Here, we define FT of the *even number data* (N/2 points)

$$G(k) = \sum_{m=0}^{(N/2)-1} x(2m)W_{\frac{N}{2}}^{mk} \tag{2.30}$$

and FT of the *odd number data* (N/2 points)

$$H(k) = \sum_{m=0}^{(N/2)-1} x(2m+1)W_{\frac{N}{2}}^{mk}. \tag{2.31}$$

Then, the N-point FT can be constructed from N/2-point FTs.

$$X(k) = G(k) + W_N^k H(k) \text{ where } k = 0, 1, 2, \ldots, N-1. \tag{2.32}$$

Now, because $W_N^{kn} = e^{-i\frac{2\pi}{N}kn}$, $W_N^k = W_N^{p+N/2} = W_N^{N/2}W_N^p = e^{-i\frac{2\pi}{N}(N/2)}W_N^p = -W_N^p$ for $k \geqslant N/2$, i.e., $k = p + N/2$ where $p = 0, 1, 2, \ldots, (N/2)-1$.

Therefore,

$X(k) = G(k) + W_N^k H(k)$ if $k = 0, 1, 2, \ldots, (N/2)-1$; and

$X(k) = G(p) - W_N^p H(p)$ if $k = p + N/2$ and $p = 0, 1, 2, 3, \ldots, (N/2) - 1$. (2.33)

In the case of $N = 8$, for example, a set of equation (2.33) becomes

(1) $\{X(0) = G(0) + W_8^0 H(0), X(1) = G(1) + W_8^1 H(1), X(2) = G(2) + W_8^2 H(2),$
$X(3) = G(3) + W_8^3 H(3)\}$ for $k = 0, 1, 2, 3$; and

(2) $\{X(4) = G(0) - W_8^0 H(0), X(5) = G(0) - W_8^1 H(1), X(6) = G(2) - W_8^2 H(2),$
$X(7) = G(2) - W_8^3 H(3)\}$ for $k = 4, 5, 6, 7$ or $p = 0, 1, 2, 3$.

In this way, we can calculate the N-point FT using N/2-point FTs. Furthermore, each of the N/2-point FTs can be calculated by $N(1/2)^2$-point FTs, and so on. FFT utilizes the computational advantage described above. Figure 2.5 illustrates this calculation step of the eight-point Fourier transform from the four-point Fourier transforms of even and odd data.

Microsoft EXCEL® has a built-in FFT function that is readily applicable without programming. However, the maximum number of points allowed for FFT is 4096. In many cases, 4096 data points with the sampling frequency 44.10 kHz (and hence the frequency resolution is 44.10 kHz/4096 = 10.77 Hz) would be sufficient for acoustic analysis. Nevertheless, if we need the frequency resolution of 5.383 Hz with the sampling frequency 44.10 kHz, we must double the data points to 8192 using the above computational technique. Refer to section 3.5 for an example of 2048-point FFT from its even and odd 1024-point FFTs.

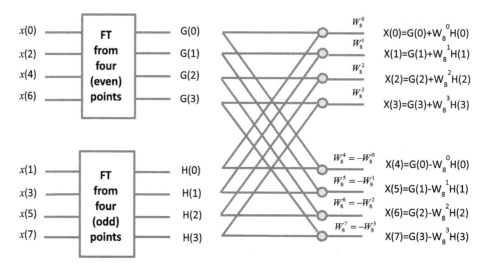

**Figure 2.5.** Eight-point discrete FT from two four-point FTs.

### 2.3.2 Algorithms of one- and two-dimensional FFTs

**One-dimensional FFT:** from the above approach, we can infer that any N-point FFT can be ultimately constructed from two-point FTs. In order to initiate the series of the two-point butterfly computations, the configurational order of the raw data, e.g. $\{x_n; n = 0, 1, 2, ..., 7\}$, needs to be changed to $\{x_0-x_4, x_2-x_6, x_1-x_5, x_3-x_7\}$. How can we obtain the correct two-point pairs when $N$ is large? This order-change can be done by 'bit-reversion' where decimal numbers are converted to corresponding binary numbers and reverse the bit order of each binary number. Table 2.1 shows the bit-reversion when $N = 8$, and figure 2.6 shows the computational path of FFT after the bit-reversion. Appendix A13.4 shows a one-dimensional FFT using EXCEL's VBA and an example of FFT using this code.

**Table 2.1.** Bit reversion of $N = 8$.

| $x$(decimal) | $x$(binary) | Reversed bit | Decima from reversed bits |
|---|---|---|---|
| 0 | 000 | 000 | 0 |
| 1 | 001 | 100 | 4 |
| 2 | 010 | 010 | 2 |
| 3 | 011 | 110 | 6 |
| 4 | 100 | 001 | 1 |
| 5 | 101 | 101 | 5 |
| 6 | 110 | 011 | 3 |
| 7 | 111 | 111 | 7 |

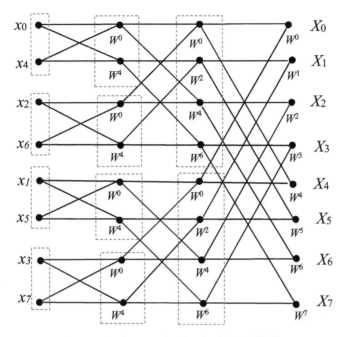

**Figure 2.6.** Computational path of N = 8 FFT.

**Two-dimensional FFT:** we need to get back to the *one-dimensional DFT* to find out the algorithm.

$$X(k) = \sum_{n=0}^{N-1} x(n)e^{-i2\pi\frac{kn}{N}} = \sum_{n=0}^{N-1} x(n)W_N^{kn}. \tag{2.34}$$

Equation (2.34) can be written in a matrix form:

$$\begin{bmatrix} X_0 \\ X_1 \\ X_2 \\ \vdots \\ X_{N-1} \end{bmatrix} = \begin{bmatrix} W_N^0 & W_N^0 & W_N^0 & \cdots & W_N^0 \\ W_N^0 & W_N^1 & W_N^2 & \cdots & W_N^{N-1} \\ W_N^0 & W_N^2 & W_N^4 & \cdots & W_N^{2(N-1)} \\ \vdots & \vdots & \vdots & & \vdots \\ W_N^0 & W_N^{N-1} & W_N^{2(N-1)} & \vdots & W_N^{(N-1)(N-1)} \end{bmatrix} \cdot \begin{bmatrix} x_0 \\ x_1 \\ x_2 \\ \vdots \\ x_{N-1} \end{bmatrix} \tag{2.35}$$

where $x_n = x(n)$ and $X_k = X(k)$ for the matric format. For simplicity, we denote the above matrix form as $[X] = [W][x]$.

The *two-dimensional DFT* can be formulated by extending one-dimensional DFT:

$$X_{\ell n} = \sum_{k=0}^{N-1}\sum_{m=0}^{N-1} x_{km}e^{-i\frac{2\pi}{N}(k\ell+mn)} = \sum_{k=0}^{N-1}\sum_{m=0}^{N-1} e^{-i\frac{2\pi}{N}\ell k}x_{km}e^{-i\frac{2\pi}{N}mn} \tag{2.36}$$

where $\ell = 0, 1, 2, \ldots, N-1$ and $n = 0, 1, 2, \ldots, N-1$. Equation (2.36) can be formulated by the following matrix form:

$$
\begin{bmatrix} X_{00} & X_{01} & \cdots & X_{0N-1} \\ X_{10} & X_{11} & \cdots & X_{1N-1} \\ \vdots & \vdots & \ddots & \vdots \\ X_{N-1,\,0} & X_{N-1,\,1} & \cdots & X_{N-1,\,N-1} \end{bmatrix} = \begin{bmatrix} W_N^0 & W_N^0 & W_N^0 & \cdots & W_N^0 \\ W_N^0 & W_N^1 & W_N^2 & \cdots & W_N^{N-1} \\ W_N^0 & W_N^2 & W_N^4 & \cdots & W_N^{2(N-1)} \\ \vdots & \vdots & \vdots & & \vdots \\ W_N^0 & W_N^{N-1} & W_N^{N-1} & \vdots & W_N^{(N-1)(N-1)} \end{bmatrix}
$$

$$
\times \begin{bmatrix} x_{00} & x_{01} & \cdots & x_{0N-1} \\ x_{10} & x_{11} & \cdots & x_{1N-1} \\ \vdots & \vdots & \ddots & \vdots \\ x_{N-1,\,0} & x_{N-1,\,1} & \cdots & x_{N-1,\,N-1} \end{bmatrix} \cdot \begin{bmatrix} W_N^0 & W_N^0 & W_N^0 & \cdots & W_N^0 \\ W_N^0 & W_N^1 & W_N^2 & \cdots & W_N^{N-1} \\ W_N^0 & W_N^2 & W_N^4 & \cdots & W_N^{2(N-1)} \\ \vdots & \vdots & \vdots & & \vdots \\ W_N^0 & W_N^{N-1} & W_N^{N-1} & \vdots & W_N^{(N-1)(N-1)} \end{bmatrix} \tag{2.37}
$$

$$
= \begin{bmatrix} X'_{00} & X'_{01} & \cdots & X'_{0N-1} \\ X'_{10} & X'_{11} & \cdots & X'_{1N-1} \\ \vdots & \vdots & \ddots & \vdots \\ X'_{N-1,\,0} & X'_{N-1,\,1} & \cdots & X'_{N-1,\,N-1} \end{bmatrix} \cdot \begin{bmatrix} W_N^0 & W_N^0 & W_N^0 & \cdots & W_N^0 \\ W_N^0 & W_N^1 & W_N^2 & \cdots & W_N^{N-1} \\ W_N^0 & W_N^2 & W_N^4 & \cdots & W_N^{2(N-1)} \\ \vdots & \vdots & \vdots & & \vdots \\ W_N^0 & W_N^{N-1} & W_N^{N-1} & \vdots & W_N^{(N-1)(N-1)} \end{bmatrix},
$$

where we define

$$
[X] = \begin{bmatrix} X_{00} & X_{01} & \cdots & X_{0N-1} \\ X_{10} & X_{11} & \cdots & X_{1N-1} \\ \vdots & \vdots & \ddots & \vdots \\ X_{N-1,\,0} & X_{N-1,\,1} & \cdots & X_{N-1,\,N-1} \end{bmatrix} \quad [x] = \begin{bmatrix} x_{00} & x_{01} & \cdots & x_{0N-1} \\ x_{10} & x_{11} & \cdots & x_{1N-1} \\ \vdots & \vdots & \ddots & \vdots \\ x_{N-1,\,0} & x_{N-1,\,1} & \cdots & x_{N-1,\,N-1} \end{bmatrix},
$$

$$
[W] = \begin{bmatrix} W_N^0 & W_N^0 & W_N^0 & \cdots & W_N^0 \\ W_N^0 & W_N^1 & W_N^2 & \cdots & W_N^{N-1} \\ W_N^0 & W_N^2 & W_N^4 & \cdots & W_N^{2(N-1)} \\ \vdots & \vdots & \vdots & & \vdots \\ W_N^0 & W_N^{N-1} & W_N^{N-1} & \vdots & W_N^{(N-1)(N-1)} \end{bmatrix}, \text{ and} \tag{2.38}
$$

$$
[X'] = [W] \cdot [x] = \begin{bmatrix} X'_{00} & X'_{01} & \cdots & X'_{0,\,N-1} \\ X'_{10} & X'_{11} & \cdots & X'_{1,\,N-1} \\ \vdots & \vdots & \ddots & \vdots \\ X'_{N-1,\,0} & X'_{N-1,\,1} & \cdots & X'_{N-1,\,N-1} \end{bmatrix}.
$$

With these definitions, the two-dimensional matrix form is converted into
$[X] = [W] \cdot [x] \cdot [W] = [X'] \cdot [W]$.

Because the transposed matrix of $[W]$ is itself. i.e., $[W]^t = [W]$, we obtain

$$[X]^t = [W]^t \cdot [X']^t = [W] \cdot [X']^t \tag{2.39}$$

by transposing matrix $[X]$. By applying this property, the two-dimensional FFT can be performed by computing two steps of one-dimensional FFTs. First, compute $[X'] = [W] \cdot [x]$ by applying one-dimensional FFT to each column:

$$
\begin{bmatrix} X'_{00} \\ X'_{10} \\ X'_{20} \\ \vdots \\ X'_{N-1,0} \end{bmatrix} = [W] \begin{bmatrix} x_{00} \\ x_{10} \\ x_{20} \\ \vdots \\ x_{N-1,0} \end{bmatrix}, \begin{bmatrix} X'_{01} \\ X'_{11} \\ X_{21} \\ \vdots \\ X'_{N-1,1} \end{bmatrix}
$$

$$
= [W] \begin{bmatrix} x_{01} \\ x_{11} \\ x_{21} \\ \vdots \\ x_{N-1,1} \end{bmatrix}, \cdots\cdots, \begin{bmatrix} X'_{0N-1} \\ X'_{1N-1} \\ X_{2N-1} \\ \vdots \\ X'_{N-1,N-11} \end{bmatrix} = [W] \begin{bmatrix} x_{0N-1} \\ x_{1N-1} \\ x_{2N-1} \\ \vdots \\ x_{N-1,N-1} \end{bmatrix}. \tag{2.40}
$$

Second, once $[X']$ is constructed, transpose the matrix $[X']$ to obtain $[X']^t$, and then compute $[X]^t = [W] \cdot [X']^t$ per column of $[X']^t$ in a similar way:

$$
[W] \begin{bmatrix} X'_{00} \\ X'_{01} \\ X'_{20} \\ \vdots \\ X'_{0,N-1} \end{bmatrix}, [W] \begin{bmatrix} X'_{10} \\ X'_{11} \\ X'_{12} \\ \vdots \\ X'_{1,N-1} \end{bmatrix}, \cdots\cdots, [W] \begin{bmatrix} X'_{N-1,0} \\ X'_{N-1,1} \\ X'_{N-1,2} \\ \vdots \\ X'_{N-1,N-1} \end{bmatrix}. \tag{2.41}
$$

Finally, we obtain $[X]$ by transposing $[X]^t$. Appendix A13.5 shows a VBA code of two-dimensional FFT and an example of the FFT.

### 2.3.3 Energy spectrum and power spectrum

Because $|x(t)|^2$ is proportional to the energy of a signal $x(t)$, the total signal energy of the signal can be given by $\int_{-\infty}^{+\infty} |x(t)|^2 dt$. By Parseval's theorem (section 2.1.2), $\int_{-\infty}^{+\infty} |X(\omega)|^2 d\omega$ is the total energy across the frequency. Because the absolute value of the Fourier component, $|X(\omega)|$, of a signal $x(t)$ corresponds to the amplitude of the signal of angular frequency $\omega$, $|X(\omega)|^2$ is the intensity (or energy). For this reason, $\{|X(\omega)|\}$ is called the amplitude spectrum, and $\{|X(\omega)|^2\}$ is called the energy spectrum density. For numerical computation of the absolute values of $\{|X(\omega)|\}$ from complex values $\{X(\omega)\}$, we can apply EXCEL's mathematical function, IMABS(). The energy spectrum can be defined only in a finite interval, e.g. $[-T/2, +T/2]$. If we have an infinitely lasting non-periodic signal, we cannot define

the energy spectrum by setting $T \to \infty$. In this case, we can consider the power spectrum density as the energy per unit of time:

$$\lim \frac{1}{T} \int_{-T/2}^{+T/2} |x(t)|^2 \, dt = \lim \int_{-\infty}^{+\infty} \frac{|X(\omega)|}{T} \, d\omega. \tag{2.42}$$

**Note:** for a stochastic process, we need to consider the expectation value of the above equation, and define the power spectral density

power spectral density =

$$\lim \left\langle \int_{-\infty}^{+\infty} \frac{|X(\omega)|}{T} \, d\omega \right\rangle. \tag{2.43}$$

## 2.4 Linear response theory

### 2.4.1 Linear black box and frequency transfer function

Let us describe the somewhat abstract idea of a linear black box [5]. As shown in figure 2.7, a linear black box is a physical system such as an acoustic instrument, an electric circuit, and an optical device such as a single slit and a lens. It may be expressed by a linear (differential) equation that describes physical phenomena.

In figure 2.7, $x(t)$ is the input signal, and $y(t)$ is the output signal or the response to the input. For a linear black box, if the input signal is a linear combination of several components $\{x_j(t); j = 1, 2, \ldots, n\}$,

$$x(t) = \sum_{j=1}^{n} a_j x_j(t) \tag{2.44}$$

where $\{a_j; j = 1, 2, \ldots, n\}$ are arbitrary constants, then the response is given by

$$y(t) = \sum_{j=1}^{n} a_j y_j(t) \tag{2.45}$$

where $y_j(t)$ is the response signal to each input signal component $x_j(t)$.

When the input $x(t)$ to the linear black box is a sinusoidal function, its output $y(t)$ is called the frequency response. Because the Fourier transform is linear, Fourier transforms of the output function should be given by the Fourier transform of the

**Figure 2.7.** Linear black box.

input function. Suppose the Fourier transforms of the input and output functions—$x_j(t)$ and $y_j(t)$ —are respectively $X_j(\omega)$ and $Y_j(\omega)$, we obtain

$$X(\omega) = \sum_{j=1}^{n} a_j X_j(\omega) \text{ and } Y(\omega) = \sum_{j=1}^{n} a_j Y_j(\omega). \tag{2.46}$$

The individual response function $Y_j(\omega)$ is given by

$$Y_j(\omega) = G(\omega)X_i(\omega) \tag{2.47}$$

where $G(\omega)$ is called the frequency transfer function characterized by the linear black box under consideration. Notice that the frequency transfer function is independent of the input signal. Using equations (2.46) and (2.47), the response function $Y(\omega)$ is given by

$$Y(\omega) = \sum_{j=1}^{n} a_j G(\omega)X_i(\omega) = G(\omega)\sum_{j=1}^{n} a_j X_j(\omega) = G(\omega)X(\omega). \tag{2.48}$$

Equation (2.48) implies that, if we know the frequency transfer function $G(\omega)$, the output response $Y(\omega)$ to an arbitrary input signal $X(\omega)$ can be calculated. Then, because the response $Y(\omega)$ is the Fourier transform of the output $y(t)$, $Y(\omega)$ can be obtained from $x(t)$ through its Fourier transform, $X(\omega)$, and the frequency transfer function, $G(\omega)$ in the following order: $x(t) \rightarrow X(\omega) \rightarrow Y(\omega) = G(\omega)X(\omega) \rightarrow y(t)$ where

$$X(\omega) = \int_{-\infty}^{\infty} x(t)e^{-i\omega t}dt \Rightarrow Y(\omega) = G(\omega)X(\omega)$$

$$y(t) = \frac{1}{2\pi}\int_{-\infty}^{\infty} Y(\omega)e^{i\omega t}dt = \frac{1}{2\pi}\int_{-\infty}^{\infty} G(\omega)X(\omega)e^{i\omega t}dt. \tag{2.49}$$

**Example 1:** suppose the input $x(t)$ is a single sinusoidal function of angular frequency $\omega_0$, i.e. $x(t) = e^{i\omega_0 t}$. Its Fourier transform becomes the Dirac delta function:

$$X(\omega) = \int_{-\infty}^{+\infty} e^{i(\omega_0-\omega)t}dt = 2\pi\delta(\omega_0 - \omega). \tag{2.50}$$

Using equation (2.49) for the single frequency input, the output $y(t)$ is given by

$$y(t) = \int_{-\infty}^{+\infty} G(\omega)\delta(\omega_0 - \omega)e^{i\omega t}dt = G(\omega_0)e^{i\omega_0 t} = |G(\omega_0)|e^{i(\omega_0 t+\varphi)} \tag{2.51}$$

where we denoted $G(\omega_0) = |G(\omega_0)|e^{i\varphi}$. The output has the same frequency as the input signal and the magnitude and phase change are given by $|G(\omega_0)|$ and $\varphi$, respectively. The impedance of electric circuits can be derived from the frequency transfer function; we will discuss this in chapter 3.

**Example 2:** when the input is an impulse—a short intense input signal at a particular time $t'$—we may use the delta function to model the impulse: $x(t) = \delta(t-t')$. In this case,

$$X(\omega) = \int_{-\infty}^{\infty} \delta(t - t')e^{-i\omega t}dt = e^{-i\omega t'}, \tag{2.52}$$

and

$$y(t) = \frac{1}{2\pi} \int_{-\infty}^{\infty} G(\omega)e^{i\omega(t-t')}dt. \tag{2.53}$$

From the definition of the inverse Fourier transform of $G(\omega)$

$$g(t) = \frac{1}{2\pi} \int_{-\infty}^{\infty} G(\omega)e^{i\omega t}dt, \tag{2.54}$$

we can obtain $y(t) = g(t-t')$ if the input is an impulse.

Equation (2.54) shows that the Fourier transform of the delta function $\delta(t)$ is 'white' in the frequency domain. Thus, we can intuitively presume that the impulse response is a superposition of responses of all frequencies with equal weight, each of which is the frequency transfer function $G(\omega)$. For this reason, the inverse Fourier transform of the frequency transfer function $g(t)$ is called the impulse response. When an impulse or the delta function-like signal is applied at time $t'$ to a linear system, the impulse response $g(t-t')$ at a later time $(t > t')$ represents the output at $t$. From equation (2.54), the frequency transfer function is given by Fourier transform of the impulse response:

$$G(\omega) = \int_{-\infty}^{+\infty} g(t)e^{-i\omega t}dt. \tag{2.55}$$

Notice that, for a physically attainable system, because of the causality, there should be no output prior to the input impulse, and thus $g(t-t') = 0$ if $t-t' < 0$. Thus, equation (2.55) should also satisfy $G(\omega) = \int_{0}^{+\infty} g(t)e^{-i\omega t}dt$.

### 2.4.2 Response to an arbitrary input function

Suppose the input is an arbitrary linear input function $f(t)$, which can be regarded as consisting of a superposition of impulses. Figure 2.8 depicts a schematic diagram of the linear response from an impulse-train where each impulse has height $f(t')$ and small width $dt'$ at $t'$. The responses to $f(t)$ is the superposition of the responses to the impulse-train $\{f(t')dt'\}$.

Because a linear response system is time invariant, the response $y(t)$ to the input $x$ $(t)$ is the same regardless of the time that the input is applied. Thus, we can acquire the response $y(t)$ at time $t$ to the impulse-train by superposing the impulse response to each impulse $f(t')dt'\delta(t-t')$ at $t'$ where $t > t'$. Adding all impulses, the net response becomes

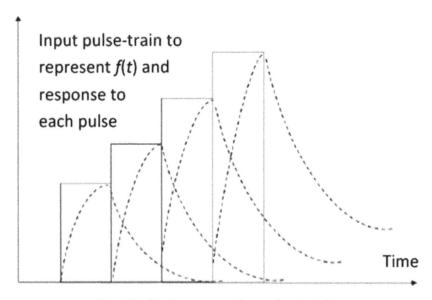

**Figure 2.8.** The linear response from an impulse-train.

$$y(t) = \sum_{\{t'\}} f(t')\delta(t - t')dt' \Rightarrow y(t) = \int_{-\infty}^{t} g(t - t')f(t')dt' \tag{2.56}$$

where $\{t'\}$ represents the time an impulse is applied, and the $g(t-t')$ is the impulse response at a later time $t$.

### 2.4.3 Harmonic oscillator with external driving term

Consider a harmonic oscillator with a damping term and an external driving term. The displacement of mass attached to the harmonic oscillator, $y(t)$, is considered to be the response to the external driving force $f(t)$ that is considered to be the input to the system. The differential equation of the harmonic oscillator is given by

$$M\frac{d^2y(t)}{dt^2} + M\omega_0^2 y(t) + M\Gamma\frac{dy(t)}{dt} = f(t) \tag{2.57}$$

where $M$, $\omega_0$, and $\Gamma$ are the mass of a particle attached to the spring, the natural oscillation frequency, and the damping factor, respectively. The third term on the left-hand-side is the damping term, and the right-hand-side, $f(t)$, represents the external driving force. Define Fourier transforms of the displacement as, $y(t)$, and the external force, $f(t)$:

$$Y(\omega) = \int_{-\infty}^{+\infty} y(t)e^{i\omega t}dt \Leftrightarrow y(t) = \frac{1}{2\pi}\int_{-\infty}^{+\infty} Y(\omega)e^{-i\omega t}dt.$$

$$F(\omega) = \int_{-\infty}^{+\infty} f(t)e^{i\omega t}dt \Leftrightarrow f(t) = \frac{1}{2\pi}\int_{-\infty}^{+\infty} F(\omega)e^{-i\omega t}dt. \tag{2.58}$$

The differential equation (2.57) can be equated as a Fourier transform to obtain

$$M(\omega_0^2 - \omega^2 - i\omega\Gamma)Y(\omega) = F(\omega), \text{ and thus } Y(\omega) = \frac{F(\omega)}{M(\omega_0^2 - \omega^2 - i\omega\Gamma)}. \quad (2.59)$$

Comparing equations (2.59) and (2.48), we obtain the frequency transfer function for the forced harmonic oscillator:

$$G(\omega) = \frac{1}{M(\omega_0^2 - \omega^2 - i\omega\Gamma)} \equiv \frac{1}{i\omega Z(\omega)} \quad (2.60)$$

where $Z(\omega)$ is called impedance of the system. When the external force $f(t)$ is an impulse, i.e. $f(t) = \delta(t-t')$, its Fourier transform is given by

$$F_\delta(\omega) = \int_{-\infty}^{+\infty} \delta(t - t')e^{i\omega t}dt = e^{i\omega t'} \quad (2.61)$$

and the impulse response is given by the inverse Fourier transform of $G(\omega)$

$$g(t - t') = \frac{1}{2\pi} \int_{-\infty}^{\infty} G(\omega)e^{-i\omega(t-t')}d\omega = \frac{1}{2\pi} \int_{-\infty}^{\infty} \frac{e^{-i\omega(t-t')}}{i\omega Z(\omega)}d\omega. \quad (2.62)$$

The response to an arbitrary driving force can be computed from

$$y(t) = \int_{-\infty}^{\infty} g(t - t')f(t')dt'. \quad (2.63)$$

**Note:** the causality requires that $g(t-t')$ must vanish when $t < t'$. Let us check if the causality is implemented in equation (2.63). Equation (2.59) can be rewritten as

$$Y(\omega) = \lim_{\varepsilon \to 0^+} \frac{1}{i\omega - \varepsilon} \frac{F(\omega)}{Z(\omega)}, \quad (2.64)$$

and hence

$$g(t - t') = \lim_{\varepsilon \to 0^+} \frac{1}{2\pi} \int_{-\infty}^{\infty} \frac{1}{i\omega - \varepsilon} \frac{e^{-i\omega(t-t')}}{i\omega Z(\omega)}d\omega. \quad (2.65)$$

If we define

$$U(\omega) = \lim_{\varepsilon \to 0^+} \frac{1}{i\omega - \varepsilon}, \quad (2.66)$$

then the inverse Fourier transform of $U(\omega)$—denoted as $u(t)$—can be calculated

$$u(t) = \frac{1}{2\pi} \int_{-\infty}^{\infty} U(\omega)e^{-i\omega t}d\omega = \frac{1}{2\pi} \lim_{\varepsilon \to 0^+} \int_{-\infty}^{\infty} \frac{e^{-i\omega t}}{i\omega - \varepsilon}d\omega = 0 \text{ if } t < 0, \text{ and } -1 \text{ if } t > 0. \quad (2.67)$$

Therefore, the causality is satisfied. Refer to appendix A3 for the detailed calculation of equation (2.65). Equation (2.67) implies that $u(t)$ is the step function. Using equations (2.65) and (2.66) for $U(\omega)$, the impulse response has the integral form

$$g(t - t') = \frac{1}{2\pi} \int_{-\infty}^{\infty} \frac{U(\omega)}{Z(\omega)}e^{-i\omega(t-t')}d\omega. \quad (2.68)$$

By the convolution theorem, equation (2.68) becomes

$$g(t - t') = \int_{-\infty}^{\infty} z(t')u(t - t')dt' \tag{2.69}$$

where $z(t) \equiv \frac{1}{2\pi} \int_{-\infty}^{\infty} \frac{e^{-i\omega t}}{Z(\omega)} d\omega$, and $g(t-t')$ vanishes if $t-t'<0$ due to the property of the step function, $u(t-t')$.

The explicit form of the impulse response of the forced harmonic oscillator is given by

$$
\begin{aligned}
g(t - t') &= \frac{1}{2\pi} \int_{-\infty}^{\infty} G(\omega)e^{-i\omega(t-t')} d\omega \\
&= \frac{i}{M} \frac{1}{\sqrt{4\omega_0^2 - \Gamma^2}} \left[ e^{-\frac{1}{2}(i\sqrt{4\omega_0^2-\Gamma^2}+\Gamma)\tau} - e^{+\frac{1}{2}(i\sqrt{4\omega_0^2-\Gamma^2}-\Gamma)\tau} \right] \\
&= \frac{i}{M} \frac{1}{\sqrt{4\omega_0^2 - \Gamma^2}} \left[ e^{-\frac{1}{2}(\Gamma+i\sqrt{4\omega_0^2-\Gamma^2})\tau} - e^{-\frac{1}{2}(\Gamma-i\sqrt{4\omega_0^2-\Gamma^2})\tau} \right]
\end{aligned}
\tag{2.70}
$$

where $\tau = t-t'$ and the transfer function $G(\omega)$ is given by equation (2.60). The detailed derivation of equation (2.70) is found in appendix A4.

### 2.4.4 Dispersion relation

If we consider a complex input function $f(t) = f_0 e^{i\omega t}$, the response function—denoted as $r(t)$—is given by

$$r(t) = \int_{-\infty}^{t} g(t - t')f(t')dt' = f_0 e^{i\omega t} \int_{-\infty}^{t} g(t - t')e^{-i\omega(t-t')}dt' = f(t)\left[ \int_{0}^{\infty} g(\tau)e^{-i\omega\tau}d\tau \right] \tag{2.71}$$

where we used equation (2.63) and $\tau = t-t'$. The integral of $g(\tau)$ appeared in equation (2.71) is called a 'Laplace-Fourier' transform. In physics, the integral

$$\chi(\omega) = \int_{0}^{\infty} g(\tau)e^{-i\omega\tau}d\tau \equiv \chi'(\omega) + i\chi''(\omega) \tag{2.72}$$

is called admittance (or susceptibility), and the complex response function to $f(t) = f_0 e^{i\omega t}$ is given by $r(t) = \chi(\omega)f(t)$.

Using equation (2.70), the explicit form of equation (2.72) for the forced harmonic oscillator is given by

$$
\begin{aligned}
\chi(\omega) &= \int_{0}^{\infty} g(\tau)e^{-i\omega\tau}d\tau = \frac{i}{M} \frac{1}{\sqrt{4\omega_0^2 - \Gamma^2}} \int_{0}^{\infty} \left[ e^{-\frac{1}{2}(\Gamma+i\sqrt{4\omega_0^2-\Gamma^2})\tau} - e^{-\frac{1}{2}(\Gamma-i\sqrt{4\omega_0^2-\Gamma^2})\tau} \right]e^{-i\omega\tau}d\tau \\
&= \frac{i}{M} \frac{1}{\sqrt{4\omega_0^2 - \Gamma^2}} \left[ \frac{1}{\frac{1}{2}(\Gamma + i\sqrt{4\omega_0^2 - \Gamma^2} + 2i\omega)} - \frac{1}{\frac{1}{2}(\Gamma - i\sqrt{4\omega_0^2 - \Gamma^2} + 2i\omega)} \right] \\
&= \frac{1}{M(\omega^2 - \omega_0^2 - i\omega\Gamma)} = \frac{1}{M} \frac{\omega_0^2 - \omega^2}{\left(\omega_0^2 - \omega^2\right)^2 + \omega^2\Gamma^2} + \frac{i}{M} \frac{\omega\Gamma}{\left(\omega_0^2 - \omega^2\right)^2 + \omega^2\Gamma^2}.
\end{aligned}
\tag{2.73}
$$

Therefore, the real and the imaginary parts of the admittance are

$$\chi'(\omega) = \frac{1}{M} \frac{\omega_0^2 - \omega^2}{(\omega_0^2 - \omega^2)^2 + \omega^2 \Gamma^2} \text{ and } \chi'(\omega) = \frac{1}{M} \frac{\omega \Gamma}{(\omega_0^2 - \omega^2)^2 + \omega^2 \Gamma^2}. \quad (2.74)$$

Notice that the real and imaginary parts of the complex admittance $\chi(\omega)$ are not independent but related with what we call the dispersion relation:

$$\chi'(\omega) = \frac{2}{\pi} \int_0^\infty \frac{\omega' \chi''(\omega')}{\omega'^2 - \omega^2} d\omega' \text{ and } \chi''(\omega) = -\frac{2}{\pi} \int_0^\infty \frac{\omega \chi'(\omega')}{\omega'^2 - \omega^2} d\omega'. \quad (2.75)$$

The dispersion relation (2.75) is also called the Kramers–Krönig relation which, for example, characterizes dielectric materials [6]. As shown in appendix A5, the dispersion relation is a consequence of the Hilbert transform:

$$\text{Re}[f(\omega)] = \frac{1}{\pi} \int_{-\infty}^\infty \frac{\text{Im}[f(\omega')]}{\omega' - \omega} d\omega' \text{ and } \text{Im}[f(\omega)] = -\frac{1}{\pi} \int_{-\infty}^\infty \frac{\text{Re}[f(\omega')]}{\omega' - \omega} d\omega'. \quad (2.76)$$

For a math exercise, readers should confirm the dispersion relation (2.75) using equations (2.74) and (2.76). The computation is lengthy but straightforward as shown in appendix A6.

## 2.5 Cepstrum

The human vocal cord is a sound source that does not produce many different forms of sound waves by itself. We produce various voice patterns by using the time-varying structure of the vocal tract consisting of various oral tissues. The vibrational sound generated by the vocal cord corresponding to the input $x(t)$ goes through the vocal tract corresponding to the impulse response $g(t)$ to output a particular voice pattern $y(t)$.

Cepstrum analysis has been used to differentiate contributions from the vocal cord and the vocal tract structure [7, 8]. Suppose $Y(k)$ is a discrete Fourier transform of a discrete time signal $y(t)$ of length $N$:

$$Y(k) = \sum_{n=0}^{N-1} y(n) e^{-i2\pi \frac{kn}{N}}. \quad (2.77)$$

We define cepstrum $c_y(m)$ of the signal $y(t)$ as the *inverse* Fourier transform of the logarithmic amplitude of $Y(k)$.

$$c_y(m) = \frac{1}{N} \sum_{k=0}^{N-1} \log |Y(k)| e^{i2\pi \frac{mk}{N}} \quad (2.78)$$

where $m$ is quefrency. Conversely, the logarithmic amplitude, $\log |Y(k)|$ can be expressed by the Fourier transform of cepstrum

$$\log |Y(k)| = \sum_{m=0}^{N-1} c_y(m) e^{-i2\pi \frac{mk}{N}}. \quad (2.79)$$

If the signal $y(t)$ is a convolution of the input $x(t)$ and the impulse response $g(t)$, we obtain $Y(\omega) = G(\omega)X(\omega)$ where $G(\omega)$ and $X(\omega)$ are Fourier transforms of $g(t)$ and $x(t)$. In such a case, equation (2.78) becomes

$$c_y(m) = \frac{1}{N}\sum_{k=0}^{N-1}\log|G(k)X(k)|e^{i2\pi\frac{mk}{N}} = \frac{1}{N}\sum_{k=0}^{N-1}(\log|G(k)| + \log|X(k)|)e^{i2\pi\frac{mk}{N}}$$

$$= c_g(m) + c_x(m)$$

(2.80)

where $c_g(m)$ and $c_x(m)$ are cepstra of $g(t)$ and $x(t)$:

$$c_g(m) = \frac{1}{N}\sum_{k=0}^{N-1}\log|G(k)|e^{i2\pi\frac{mk}{N}},$$

$$c_x(m) = \frac{1}{N}\sum_{k=0}^{N-1}\log|X(k)|e^{i2\pi\frac{mk}{N}}.$$

(2.81)

The above equations imply that the impulse response and the input can be separated if $c_g(m)$ and $c_x(m)$ are well separated. In chapter 3, we discuss how to separate the vocal tract structure from the vocal cord vibration.

# References

[1] Batenkov D 2005 *Fast Fourier transform* (Key Papers in Computer Science Seminar) 2005 http://www.wisdom.weizmann.ac.il/~naor/COURSE/fft-lecture.pdf

[2] *Nyquist Theory* Wolfman MathWorks® (http://mathworld.wolfram.com/NyquistFrequency. html).

[3] Cooley J W and Tukey J W 1965 *An algorithm for the machine calculation of complex Fourier series Math. Comput.* **19** 2.97301297E8–301

[4] *Fourier Transforms and the Fast Fourier Transform (FFT) Algorithm* (http://cs.cmu.edu/afs/ andrew/scs/cs/15–463/2001/pub/www/notes/fourier/fourier.pdf).

[5] Tokmakoff A 2009 *Linear Response Theory* (MIT Department of Chemistry) https://home. uchicago.edu›~tokmakoff›TDQMS›Notes›8...

[6] Wolfram MathWorld 2023 *Dispersion Relation* (Dispersion Relation—from Wolfram MathWorld)

[7] Oppenheim A V and Schafer R W 2004 *From frequency to quefrency: a history of the cepstrum IEEE Signal Process Mag.* **21** 5

[8] *LECT 13: Cepstral Analysis—ISIP* (https://isip.piconepress.com/.../current/lecture_13/lec-ture_13.pdf).

**IOP** Publishing

Fourier Transform and Its Applications Using Microsoft EXCEL®
(Second Edition)

Shinil Cho

# Chapter 3

# Hands-on Fourier transform using EXCEL®

In this chapter, we demonstrate practical Fourier transform (FT) computational skills, such as windowing functions, peak peeking calculations, inverse FT, and a $N$-point fast Fourier transform (FFT) from two ($N/2$)-point FFTs. We also describe acoustic spectrum analysis with hands-on FT experiments on human voices and musical instruments.

For the hands-on experiments on Fourier transform, a notebook computer is used to acquire and analyze various acoustic data. For performing a Fourier transform, the built-in data analysis and chart function of Microsoft EXCEL® are utilized. With EXCEL's built-in FFT capability, computational routines can be simplified without additional programming, and the whole process from data acquisition to spectrum analysis can be carried out with a personal computer even at home.

This chapter is written for those who never tried to compute a Fourier transform using experimental data. This chapter can be skipped if readers are mainly interested in theoretical aspects of Fourier transform.

## 3.1 Data acquisition

**Hardware**: a personal computer with a microphone is the only hardware needed for acoustic spectrum analysis. A built-in microphone serves well for acquiring sound data, and PCs with various generations of CPU and Windows operating systems can be used without any problem.

**Software**: for acquiring sound signals, any sound recording and editing software may be used. Ideally, such recording software should be able to adjust the sampling frequency and the input gain. Prior to carrying out a FT, the acquired sound data is displayed in order to determine the start and end times (and the sample size) for the FT. The sound data can be acquired in the waveform audio file (WAV) format where the maximum sampling frequency is set to 44.10 kHz. According to the

Nyquist theorem described in section 2.2.1, the maximum detectable sound frequency will be 22.05 kHz when the sampling frequency is 44.10 kHz. This setting fully covers the audible sound frequency range.

Any version of Microsoft EXCEL® has the FT routine. However, each version of EXCEL has different steps to the FT routine. This book uses EXCEL 2019, and readers who use the version 2007 or older, should be aware of the location of the data analysis option in the menu.

It should be noted that the FT analysis of this book is not in real-time. The scheme does not display a spectrum as an FT is performed. However, it is advantageous for studying the FT for the first time because readers can confirm and visualize each step of acquiring from a raw signal to its Fourier spectrum and understand the various practical issues and techniques.

**Flow chart:** Figure 3.1 is the flow chart for acquiring sound spectrum.

1) **Data acquisition**: the data acquisition is straightforward: we run sound recording software [1] to record the sound of a human voice or musical instrument. Care should be taken to avoid the saturation of sound signals. If the intensity of the source sound is excessive, the gain reduction should be carried out to avoid saturation. For data acquisition, the noise suppression function of the sound recording software should be disabled to avoid DC offset and loss of weak signals. Acquired sound data can be displayed on screen using a sound editor software [2] to determine the starting time of FFT and the number of data points (and the ending time).

2) **Data format conversion**: unfortunately, EXCEL does not read data in the WAV format, and we need to convert data from the WAV format to the comma separated value (CSV) format. There are several conversion programs available on the internet [3], although some of them are incompatible with the latest Windows operating system.

**Remark:** some format conversion software may not automatically add the file extension '.CSV' to a converted data file even if they properly perform the format conversion. If this occurs, manually adding the file extension '.CSV' will have EXCEL recognize the converted CSV file properly (figure 3.1).

Figure 3.2 shows the beginning part of a converted CSV file input in an EXCEL sheet. In this file, the data in column A show the real-time sequence, and cell A1 shows the starting time. The data in column B are the sound signal values.

**Note:** EXCEL does not save calculation results—including the FT—in the CSV format. Save the created file in the *Workbook* format as soon as possible. If a PC has a stereo speaker, there will be another data set in column C but we only need monaural data. The average value for each cell can be calculated by inputting = **(B1+C1)/2** in cell D1 and then applying the *autofill* function to the entire column D. In this case, the FT will be in column F.

Once we confirm that EXCEL has correctly read the signal data, we input the data in sequential numbers ($k$; $k$ = 0, 1, 2, ..., $N-1$) in column C to calculate the frequency step. First, enter **0** in cell C1, and then enter = **1+C1** in cell C2. Second, use the *autofill* function to complete the formulas in column C. Click the lower right

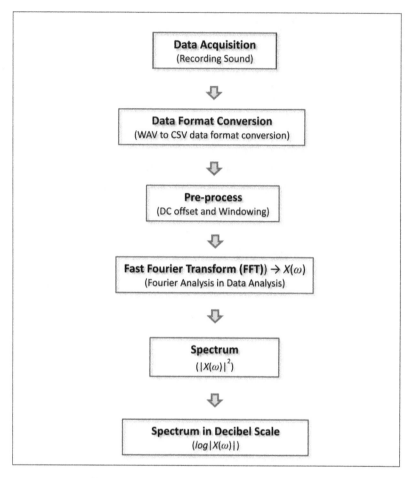

**Figure 3.1.** Flow chart of acoustic sound spectrum.

**Figure 3.2.** EXCEL sheet of acquired data in CSV format.

| | A | B | C | D | E | F |
|---|---|---|---|---|---|---|
| 1 | 1.002018 | 6130 | 0 | 0 | | |
| 2 | 1.002041 | 4952 | 1 | 10.77 | | |
| | | 35 | 2 | 21.54 | | |
| | | | 3 | 32.31 | | |
| | | 155 | 4 | 43.08 | | |
| | | 405 | 5 | 53.85 | | |
| | | -113 | 6 | 64.62 | | |
| | | -572 | 7 | 75.39 | | |
| 9 | 1.0022 | -1206 | 8 | 86.16 | | |
| 10 | 1.002222 | -2035 | 9 | 96.93 | | |

1. Enter 0 in Cell C1.
2. Enter =C1+1 in Cell C2. Then apply autofill to complete the data.

Enter =10.77*A1 in Cell D1. Then use autofill to complete the frequency step.

**Figure 3.3.** EXCEL sheet completing frequency step. **Note:** if average values of a stereo sound signal are in column D, the frequency step will be in column E and the FT will be in column F.

corner of cell D1 and drag the mouse pointer to cell D4096 if a 4096-point FT is to be calculated. Refer to appendix A12.1 for the *autofill* function.

With the data in sequential numbers created in column C, the frequency step (or resolution) for spectrum data can be easily created in column D. The number of data points for the FT and the sampling frequency determine the sampling time interval and frequency resolution of the spectrum. If the sampling frequency is $f_c$ (Hz) and the number of data points is $N$, the frequency increment of the FT spectrum will be $f_c/N$ (Hz). For example, in the data set of figure 3.2, the number of data points is 4096 and the sampling frequency is 44.10 kHz. The frequency step is given by (44.10)(1000)/4096 = 10.766 ≈ 10.77 Hz. Enter = **10.77*C1** in Cell D1 and apply the *autofill* function to make the frequency step. Figure 3.3 illustrates this step.

## 3.2 Computational steps to perform EXCEL's Fourier transform

We are now ready to compute a FT. Table 3.1 is a lookup table of the frequency increment of typical sound data points and sampling frequencies.

Figure 3.4 shows the EXCEL sheet after generating the frequency increments in column D. If preferred, the data number and the frequency step can be obtained simultaneously by applying the *autofill* function to columns C and D simultaneously.

1. **Pre-process (DC offset and windowing):** prior to FFT, we carry out the DC offset and a windowing process as required. As we pointed out in section 2.2.3, a truncated measurement produces abrupt signal changes at both ends of the sampled time interval, causing an artificial ghost around the spectral peak. In order to avoid the unwanted signal, we convolute a windowing function (a weight function) to make the signal gradually zero at the ends. Table 3.2 lists popular windowing functions.

First, for the DC offset, calculate the average value of the 4096 data points (column B) using the AVERAGE function. Second, to apply the Hanning windowing, enter = **(B1-[average value])*(1-COS(2*PI()*C1/4095))** in cell E1 where we also include the DC offset. Use the *autofill* function to calculate windowed data value in column E as shown in figure 3.4. Refer to table 3.2 for other windowing functions.

**Table 3.1.** FT parameter for sound acquisition.

| # Data points for FFT ($N$) | Frequency resolution Sampling time interval | Sampling frequency $f_c = 44.10$ kHz | Sampling frequency $f_c = 22.05$ kHz | Sampling frequency $f_c = 11.025$ kHz |
|---|---|---|---|---|
| 1024 | Sampling time duration (msec) | 23.22 | 46.44 | 90.70 |
| | Frequency resolution: $f_c/N$ (Hz) | 43.07 | 21.54 | 10.77 |
| 2048 | Sampling time duration (msec) | 46.44 | 90.70 | 181.4 |
| | Frequency resolution: $f_c/N$ (Hz) | 21.54 | 10.77 | 5.383 |
| 4096 | Sampling time duration (msec) | 90.70 | 181.4 | 362.8 |
| | Frequency resolution: $f_c/N$ (Hz) | 10.77 | 5.383 | 2.693 |
| 8192* | Sampling time duration (msec) | 181.4 | 362.8 | 725.6 |
| | Frequency resolution: $f_c/N$ (Hz) | 5.383 | 2.693 | 1.346 |

Notes[1] 8192 point FFT can be constructed by using 4096-point FFTs of the even and the odd terms (section 2.3.1).

**Figure 3.4.** Pre-process of DC offset and Hanning windowing. **Note**: we may use the *autofill* function to calculate columns C, D, and E at once.

**Table 3.2.** Common windowing functions.

| Window | Function Form ($n = 0, 1, 2, ..., N-1$) |
|---|---|
| Hanning | $m - (1-m)\cos\left(\frac{2\pi n}{N-1}\right)$ where $m = 0.5$ |
| Hamming | $m - (1-m)\cos\left(\frac{2\pi n}{N-1}\right)$ where $m = 0.54$ |
| Blackman | $0.423 - 0.498\cos\left(\frac{2\pi n}{N-1}\right) + 0.0792\cos\left(\frac{2\pi \cdot 2n}{N-1}\right)$ |
| Blackman–Harris | $0.35875 - 0.48829\cos\left(\frac{2\pi n}{N-1}\right) + 0.14128\cos\left(\frac{2\pi \cdot 2n}{N-1}\right) - 0.01168\cos\left(\frac{2\pi \cdot 3n}{N-1}\right)$ |

**Note:** there are other known windowing functions but those given in table 3.2 are the most commonly used windowing functions.

**Figure 3.5.** Data analysis in [Data] menu.

2. **FFT**: to enable EXCEL to perform a FFT, we need to turn on *Data Analysis* in EXCEL. The default setting of EXCEL does not have it, and we need to install it. Refer to appendix A12.2 for setting up this one-time option.

Once the data analysis is enabled, click on the menu [Data] and then click on [Data Analysis] which appears on the right side of the menu. The [Data Analysis] menu and then [Data Analysis Tools] will be displayed as shown in figures 3.5 and 3.6:

Click on [Fourier Analysis] to display its data selection menu and specify the parameters for a FFT as shown in figure 3.7. Click on the [Input Range] box and enter $E$1:$E$4096. Select [Output Range] to obtain the FFT on the same sheet. Click on the [Output Range] box and enter $F$1:$F$4096 to have the computed FFT in column F. Click on [OK] to display the FT data in column F.

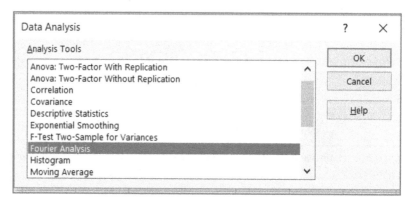

**Figure 3.6.** Data analysis tools.

1. Place the cursor in Cell E1.
2. Click this box. Excel shows $E$1:$E$1.
3. Replace the last row number 1 with $N$ (e.g., 4096) as shown.

**Figure 3.7.** Fourier analysis parameter entry.

The horizontal size of column F can be enlarged to display long complex data strings as shown in figure 3.8.

3. **Energy spectrum**: figure 3.9 illustrates the steps to obtain energy spectrum. The obtained FFT data are generally complex numbers. Apply the IMABS() function to obtain the absolute values, $|X(\omega)|$. For the energy spectrum, we need to calculate $|X(\omega)|^2$. Enter = (IMABS(F1))^2 in cell G1 and use the *autofill* function to calculate the absolute value. In column H, we calculate power in the decibel unit. Enter = 20×LOG(G1) in cell H1. We can also apply the *autofill* function for calculating all terms of column H.

| | A | B | C | D | E | F | G | H |
|---|---|---|---|---|---|---|---|---|
| 1 | 1.002018 | 6130 | 0 | 0 | 0 | -22011˙ ˙ .̣oo9 | | |
| 2 | 1.002041 | 4952 | 1 | 10.77 | 0.00582 | 11889.4625026508-1495.43⁴˙ | ˌ38i | |
| 3 | 1.002063 | 3510 | 2 | 21.54 | 0.016495 | -1078.57202683703+?˙ | ˌ955613i | |
| 4 | 1.002086 | 2164 | 3 | 32.ˌˌ˙ | Right click on this column border | | ˙˙ˌˌˌˌ˙02i | |
| 5 | 1.002109 | 1155 | 4 | 43 | and move it to the right to expand | | ᵢi | |
| 6 | 1.002132 | 405 | 5 | 53 | | | 5i | |
| 7 | 1.002154 | -113 | 6 | 64 | the cell width. | | .4i | |
| 8 | 1.002177 | -572 | 7 | 75 | | | | |
| 9 | 1.0022 | -1206 | 8 | 86.16 | -0.09118 | -10440.ɔ21ɔɔ26oɔ+2ɔ8ɔ1.ɔ448zɔ6ɔ37i | | |
| 10 | 1.002222 | -2035 | 9 | 96.93 | -0.19441 | 26694.6606243783-39237.3110201982i | | |

**Figure 3.8.** FFT data in column F.

1. Enter = `(IMABS(F1))^2` in Cell G1.

2. *Autofill* to complete Column G.

3. Enter =`20*LOG(G1)` in Cell H1.

4. *Autofill* to complete Column H.

| | A | B | C | D | E | F | G | H | I |
|---|---|---|---|---|---|---|---|---|---|
| 1 | 1.002018 | 6130 | 0 | 0 | 0 | -22011.92609 | 484524890 | 173.7063 | |
| 2 | 1.002041 | 4952 | 1 | 10.77 | 0.00582 | 11889.4625026508-1495.43415984138i | 143595642 | 163.1428 | |
| 3 | 1.002063 | 3510 | 2 | 21.54 | 0.016495 | -1078.57202683703+3612.13197955613i | 14210815 | 143.0524 | |
| 4 | 1.002086 | 2164 | 3 | 32.31 | 0.022862 | 2853.95062696549-2993.72952629702i | 17107451 | 144.6637 | |
| 5 | 1.002109 | 1155 | 4 | 43.08 | 0.021649 | -5970.3458767846+1197.54213757869i | 37079137 | 151.3826 | |
| 6 | 1.002132 | 405 | 5 | 53.85 | 0.011766 | 3005.5337278826-4931.89779867642i | 33356849 | 150.4637 | |
| 7 | 1.002154 | -113 | 6 | 64.62 | -0.005 | 4417.09611778538+4978.63135373545i | 44297508 | 152.9276 | |
| 8 | 1.002177 | -572 | 7 | 75.39 | -0.03326 | -2474.16977453476-5122.32347122014i | 32359714 | 150.2001 | |
| 9 | 1.0022 | -1206 | 8 | 86.16 | -0.09118 | -10440.521952609+23891.5448298937i | 679810413 | 176.6478 | |
| 10 | 1.002222 | -2035 | 9 | 96.93 | -0.19441 | 26694.6606243783-39237.3110201982i | 2.252E+09 | 187.052 | |

**Figure 3.9.** Energy spectrum from FT data.

4. **Graphing spectrum**: for graphing the spectra obtained above, we use a scatter chart with a smooth graph. Highlight columns D and H, and then click on the [Insert] menu. Go to [Chart] and select [Scatter with Smooth Line].

5. **Note on Nyquist theorem and side lobe**: let us follow the above steps for a pure sinusoidal wave of frequency 50 Hz, $x(t) = \sin(100\pi t)$, and create its amplitude spectrum. In this case, we select the sampling frequency 1 kHz which is well over the Nyquist frequency of this signal, 100 Hz. We take 1024 data points. The frequency resolution is thus $1000/1024 = 0.977$ Hz. Figures 3.10 and 3.11 show the signal in the time domain (the first 100 data points) and the 1024-point FFT, $\{|X(\omega)|\}$. At first, only a square windowing function (or finite-time truncation) is applied.

Notice that the second half of the FFT graph is the reversed repetition of the first half as a result of the Nyquist sampling theorem (section 2.2.1) and the FFT algorithm employed by EXCEL. In other words, the useful spectrum by FFT has $N/2$ data points, provided $N$ is the number of data points in the time domain. In this case, the meaningful data points are $1024/2 = 512$, and the total frequency range is $(0.977)(512) \approx 500$ Hz. Taking the frequency range up to 100 Hz to display the spectrum, we obtain the amplitude spectrum as shown in figure 3.12. Side lobes are noticeable as long tails because of the square windowing function.

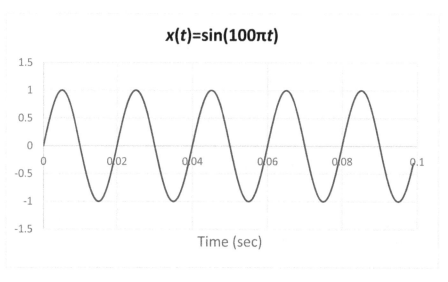

**Figure 3.10.** Single sinusoidal signal of 50 Hz.

**Figure 3.11.** FFT of pure sine function.

## 3.3 The effect of the windowing function

As we discussed in section 2.2.3, side lobes in the spectrum shown above are caused due to the finite length of data where the initial and final data points of the data do not have the same value, which generates artificial periodic noise. As pointed out, there are several windowing functions that may suppress the side lobes. Table 3.2 lists common windowing functions used for spectrum analysis [4].

**Figure 3.12.** Energy spectrum of sine wave.

**Figure 3.13.** Sinusoidal signal spectrum with Hanning windowing.

Let us use the amplitude spectrum of a pure sine function shown in figure 3.11 to observe the effect of each windowing function listed. Figures 3.13–3.16 show the spectra from data applied to several windowing functions. The Hanning window is simple and effective. Line narrowing or reduction of the side lobes is effective because of the windowing functions. While the line narrowing is similar, there is some structural difference in the tails. In this book, the Hanning window is applied for spectra acquired below when side lobes are noticeable.

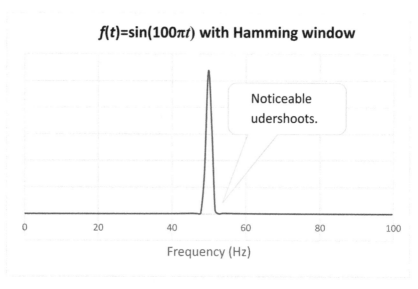

**Figure 3.14.** Sinusoidal signal spectrum with Hamming windowing.

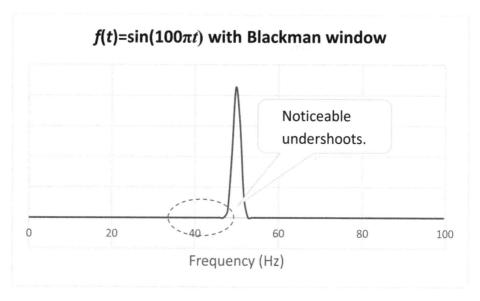

**Figure 3.15.** Sinusoidal signal spectrums with Blackman window.

## 3.4 Peak peeking

There are several methods for interpolating the peak location of an FT when the frequency of interest is between bins [5]. Since EXCEL has a built-in curve fitting routine [Trendline], we can apply it to estimate the peak frequency and the peak value.

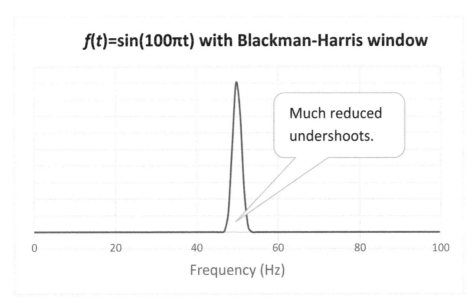

**Figure 3.16.** Sinusoidal signal spectrum with Blackman–Harris window.

For applying EXCEL's curve fitting routines in [Trendline], click on any data point of the scatter chart and then click on [Design] of the [Chart Tools] menu. Click on the [Add Chart Element] menu and select [Trendline]. Select [More Trendline Options...] and check [Polynomials]. EXCEL offers up to a sixth-order polynomial curve fitting. By checking the [Display Equation on Chart] box, the polynomial equation is displayed on the screen. Figure 3.17 shows the second order polynomial fitting.

Figure 3.18 shows a peak of the energy spectrum of a voice. The frequency resolution is 10.77 Hz (from the sampling frequency 44.10 kHz and 4096 data points). Because the frequency resolution is insufficient, we encounter a situation where we cannot determine the frequency of a spectral peak precisely.

In this case, we can find the peak position and values by applying the quadrature curve fitting using EXCEL as follows:

1. Select three data points and make a scatter graph.
2. Apply trendline [Polynomial] and [Order 2] to find the best fitted quadrature equation for the three points around a peak checking [Display Equation on Chart]. Visually check the curve fitting (figure 3.19). The equation displayed on the chart is quadrature: $y = ax^2 + bx + c$.

   **Note:** set the three frequency values to −10.77, 0, and +10.77 Hz for this curve fitting otherwise EXCEL's polynomial fitting displays incorrect quadrature equations.

   From the data of table 3.3, we obtain the quadrature equation shown below.

3. Change the equation to $y = a\left(x - \frac{b}{2a}\right)^2 + \left(c - \frac{b^2}{4a}\right)$. Then, the peak value is $c - (b^2/4a) = 234\ 910$ and the peak frequency is $(b/2a) = 2.97$ in the reduced

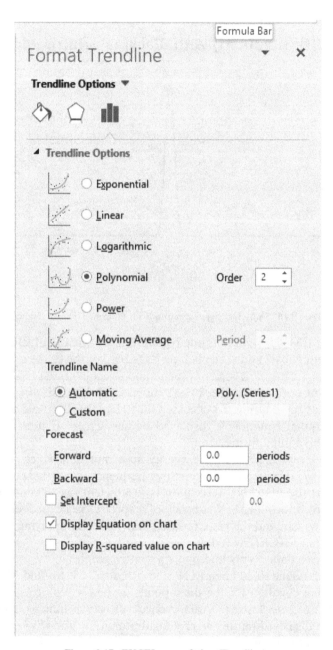

**Figure 3.17.** EXCEL curve fitting (Trendline).

frequency scale. Adding the center frequency 1184 Hz to the obtained peak frequency gives 11 847 Hz in this case.

**Remark:** we can also apply the quadrature curve fitting to the logarithmic spectrum. The reduced frequency should be used to avoid an incorrect curve fitting equation.

**Figure 3.18.** Spectrum around a peak frequency.

**Figure 3.19.** Three-point quadrature curve fitting near peak.

**Table 3.3.** EXCEL data around peak.

| $f$ (Hz) | Reduced frequency (Hz) | $|X(\omega)|$ |
|---------|------------------------|---------------|
| 1163.16 | −21.54 | 117 908.8 |
| 1773.93 | −10.77 | 119 790.6 |
| 1184.17 | 0 | 230 569.7 |
| 1195.47 | 10.77 | 191 840.1 |
| 1206.24 | 21.54 | 111 902.4 |

In this case, the peak frequency is also 11 847 Hz and the peak value is 5.38, which is converted to 239 883 in the linear scale.

## 3.5 Demonstration of *N*-point FFT from two (*N/2*)-point FFTs

The maximum number of data points with which EXCEL can perform an FT is 4096. However, we can carry out an 8192-point FFT by separating the even and the odd points of 8192 data as we described in section 2.3.1. For this computation, we need to rearrange a set of data points to its even terms and odd terms. A simple VBA macro program can perform this routine. Figure 3.20 is the amplitude spectrum of a piano sound generated by the 2048-point FFT where the fundamental mode and the first higher harmonic mode are shown.

Let us create the $N$ (= 2048)-point FFT from the even and the odd term FFTs. This demonstration uses raw data without any windowing.

After rearranging the even and odd terms, we compute FFT from each of the separated data sets. The amplitude spectra shown in figures 3.21 and 3.22 are obtained from these data points. Appendices A12.3 and A13.2 explain how to enable EXCEL's VBA macro capability and use the macro to rearrange even and odd terms.

The two 1024-point FFTs can be constructed using the following equation (3.1), which is the same as equation (2.32):

$$X_k(\omega) = G_k(\omega) + W_N^k H_k(\omega) = G_k(\omega) + e^{-i2\pi nk/N} H_k(\omega)$$
$$= G_k(\omega) + (e^{-i2\pi n/N})^k H_k(\omega) \tag{3.1}$$

where $G(k)$ and $H(k)$ are the even term FFT and the odd term FFT, respectively. The power of complex numbers by applying the IMPOWER() function, and the product of complex numbers can be obtained by the IMPRODUCT() function. Figure 3.23 below is the amplitude spectrum obtained, which is identical to figure 3.20.

## 3.6 Inverse Fourier transform

EXCEL can also perform an inverse FFT. By checking the [Inverse] box of the [Fourier Analysis] popup menu (figure 3.6). Its procedure is similar to the

**Figure 3.20.** 2048-point amplitude spectrum of piano sound (G6 = 1568 Hz).

**Figure 3.21.** 1024 even term FFT.

ordinary FFT. Consider the 4096-point FFT of the following signal of two different frequencies, 10 Hz and 60 Hz:

$$x(t) = \sin(60\pi t) + 0.5 \sin(20\pi t). \tag{3.2}$$

Prior to performing a FFT, we apply the Hanning window. Figure 3.24 is the original time-domain signal with the Hanning window. In this example, columns A, B, C, and D are data number, time, raw data, and the windowed data, respectively, as shown in figure 3.25. The FFT data and their absolute values are shown in columns E and F of the figure. Figure 3.26 shows the amplitude spectrum of the two sine waves.

**Figure 3.22.** 1024 odd term FFT.

**Figure 3.23.** $N$ FFT constructed from ($N/2$)-even and ($N/2$)-odd FFTs ($N = 2048$).

We add the inverse FFT values to column G, using the input data in column E (see figure 3.25). While the first half ($N/2$) data points are useful for the spectrum, we should use the full ($N$) data points for inverse FTs. Click on [Fourier Analysis] to display its data selection menu and check the [Inverse] box as shown in figure 3.27.

It should be noted that EXCEL may store the inverse FFT data as text. This situation is shown in figure 3.28. If this happens, convert the column text into numerical values by applying the [Text to Columns] menu in the [Data] menu as shown in figure 3.29.

**Figure 3.24.** Two sine waves of frequency 30 Hz and 10 Hz.

| | A | B | C | D | E | F |
|---|---|---|---|---|---|---|
| 1 | 0 | 0 | 0 | 0 | -0.317574071275496 | 0.317574 |
| 2 | 1 | 0.00025 | 0.05496 | 6.46E-08 | -0.334963633396191+9.09990873612...e-002i | 0.347104 |
| 3 | 2 | 0.0005 | 0.109814 | 5.17E-07 | | 0.441161 |
| 4 | 3 | 0.00075 | 0.164454 | 1.74E | 1. Enter = **IMABS (E1)** in Cell F1 | 0.623902 |
| 5 | 4 | 0.001 | 0.218777 | 4.12E | 2. *Autofill* to calculate $|X(\omega)|$ | 0.961855 |
| 6 | 5 | 0.00125 | 0.272675 | 8.02E-06 | -1.18512767293152+1.115125718157641 | 1.627278 |
| 7 | 6 | 0.0015 | 0.326045 | 1.38E-05 | -2.20446913897887+2.20626908032255i | 3.118863 |
| 8 | 7 | 0.00175 | 0.378785 | 2.18E-05 | -5.05295841493896+5.23441957740547i | 7.275406 |
| 9 | 8 | 0.002 | 0.430791 | 3.24E-05 | -17.0884962562548+18.0176584946766i | 24.83249 |
| 10 | 9 | 0.00225 | 0.481965 | 4.59E-05 | -229.335495297923+243.634380026875i | 334.593 |

**Figure 3.25.** FFT data: $X(\omega)$ in column E and $|X(\omega)|$ in column F.

**Figure 3.26.** Amplitude spectrum.

3-18

**Figure 3.27.** EXCEL's inverse FFT.

| F | G |
|---|---|
| 0.317574 | 0 |
| 0.34 | 6.46316604679953E-008 |
| 0.44 | 7 |
| 0.62 | 5 |
| 0.96 | 5 |
| 1.62 | 5 |
| 3.11 | 5 |
| 7.27 | 5 |
| 24.8 | 5 |
| 334 | 5 |
| 986.7366 | 6.25850323592303E-005 |
| 695.091 | 8.27304248988137E-005 |

Number Stored as Text

Convert to Number

Help on this Error

Ignore Error

Edit in Formula Bar

Error Checking Options...

**Figure 3.28.** Inverse FFT data stored as text.

Highlight column G and select the [Text to Column] menu from the [Data] menu. Select the [Delimited] data type and click on [Next] and then [Finish] to complete the conversion.

Figure 3.30 shows the signal obtained by the inverse FFT, which is identical to the original signal in the time domain (figure 3.24).

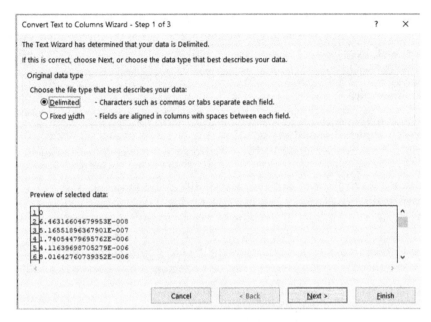

**Figure 3.29.** Text to numerical data conversion.

**Figure 3.30.** Inverse FFT of FFTed Signal.

## 3.7 Acoustic spectra

Acoustic sounds from the human vocal system and musical instruments have complicated wave forms, exhibiting different pitches and timbres. The acoustic spectrum analysis is an ideal subject to use to learn Fourier transform and pre-processes. It also provides us with much information on physical systems including

the structures of musical instruments and conditions of machine tools [1, 2]. Time-frequency analysis of timbre also gives useful information on speech therapy, sound synthesis, and recent voice recognition technology.

### 3.7.1 Human voice

Figures 3.31 and 3.32 show an example of acquired voice data using the PC system, described in section 3.1, and the initial part of the data sequence, respectively. The sampling frequency used is 44.10 kHz. The data format is converted to the WAV format. Since the sampling frequency is 44.10 kHz, the time duration should be 92.88 ms or longer in order to have at least 4096 data points.

A starting time is selected when the signal value is close to zero. For this sample, the starting time at 1.2 s, and the ending time 1.3 s to have the sampling time interval of 100 ms are appropriate. Observing the obtained full (2048 points) spectrum, the maximum voice frequency is approximately 5 kHz, and the spectrum up to 5 kHz is shown in figure 3.33. The spectrum can be normalized using a standard numerical integral method if necessary.

The sound of a human voice is generated by air flow that causes the vibration of the vocal cords. Similar to the frequency of an oscillating string, the frequency of the generated sound is approximately proportional to the tension in the vocal cords. If the vocal cords have more mass, the generated sound tends to have a lower frequency. Sounds generated by the vocal cords do not have much variation in waveform. The vibration of the vocal cords determines the pitch of the voice.

The sound from the vocal cords becomes various voice sound forms while passing through the time-varying vocal tract structure, which is a cavity consisting of the

**Figure 3.31.** Recorded voice.

**Figure 3.32.** Initial part of recorded voice.

**Figure 3.33.** Human (female) voice spectrum.

hard palate, the soft palate, uvula, pharynx, and tongue. This process of creating a particular vocal tract structure can be regarded as resonance or filtering. The peaks in the voice spectrum are resonance frequencies determined by the vocal tract. Because the vocal tract functions as a resonator or a filter, the linear response theory (section 2.4) can be applied to the human voice system. In other words, a particular voice sound pattern is produced by the convolution of the vocal cord vibration and the vocal tract resonator. The Fourier transform of output signal $Y(\omega)$ is given by the product of the Fourier transforms of source signal $X(\omega)$ from the vocal cords and vocal tract transfer function (or filter) $G(\omega)$ formed by the vocal tract structure, $Y(\omega) = G(\omega) X(\omega)$.

In order to observe the effect of the vocal tract more clearly, the linear scale is converted to the logarithmic (decibel) scale, which matches human hearing perception. With the logarithmic scale, higher harmonics are easily identified. In addition, we observe a broad structure across the frequency range of human voices. The logarithmic scale makes it easier to separate the input signal $X(\omega)$ from the vocal tract transfer function $G(\omega)$: log $Y(\omega)$ = log $G(\omega)$ + log $X(\omega)$. In figure 3.34, we attempted to fit the broad structure, corresponding to the log $G(\omega)$-term, by using a polynomial fitting of the sixth-order using EXCEL's curve fitting routine which we used for peak peeking (section 3.4).

### 3.7.2 Notes on vocal formant

The vocal tract forms the broad spectral structure, which is called a formant. The formant analysis is important in voice recognition and other technologies. It is interesting to investigate how the peak frequencies and the broad spectral structure change as one changes pitch or timbre by generating several vowels at different

**Figure 3.34.** Energy spectrum in decibel scale.

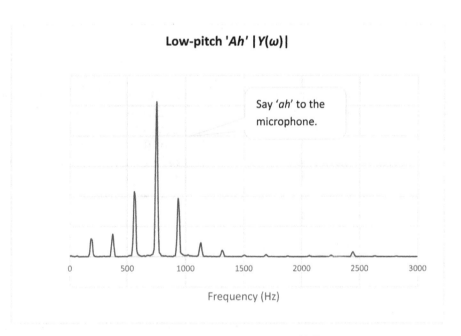

**Figure 3.35.** low-pitch (male) sound of '*ah*'.

pitches. Figures 3.35–3.38 show amplitude spectra of a low-pitch and high-pitch sounds of the vowel '*ah*'.

The spectrum of high-pitch sound appears to be two sets of sound spectra. In other words, when we produce the same vowel '*ah*' at different pitches, the vocal cords generate different frequency components while the structural shape of the vocal cavity is the same. When we hear an '*ah*' sound, we recognize the sound as '*ah-timbre*' regardless of its pitch. Since our hearing perception is logarithmic, there must

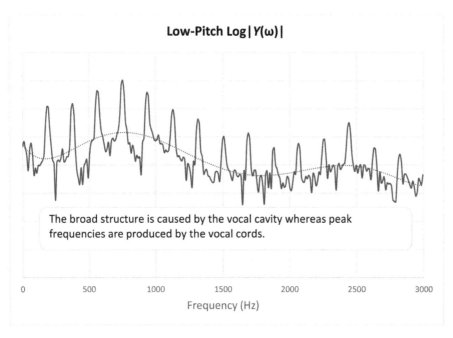

**Figure 3.36.** low-pitch (male) sound of *'ah'*.

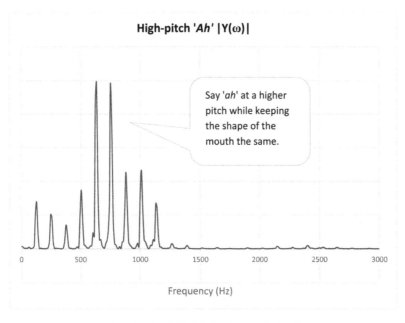

**Figure 3.37.** High-pitch (male) sound of *'ah'*.

**Figure 3.38.** High-pitch (male) sound of '*ah*'.

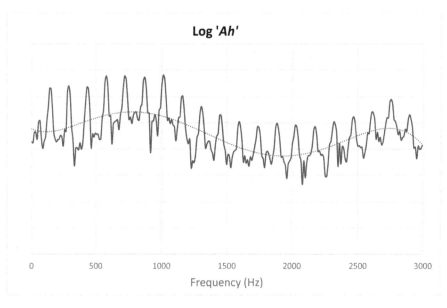

**Figure 3.39.** High-pitch (male) sound of '*ah*'.

be some structure in the spectrum that is the same for all '*ah*' sounds. That structure should be characterized the vocal tract transfer function $G(\omega)$.

Figures 3.39–3.41 are logarithmic energy spectra of vocal sounds, '*ah*', '*oh*', and '*ee*' at the same pitch. If we actually generate these sounds by ourselves, we notice that the tension in the vocal cords does not change for the same pitch and it produces a similar set of sound frequencies while we change the shape of the oral cavity and lips.

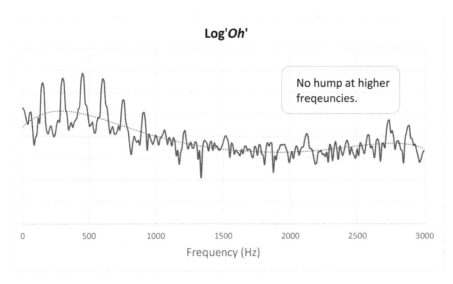

**Figure 3.40.** High-pitch (male) sound of '*oh*'.

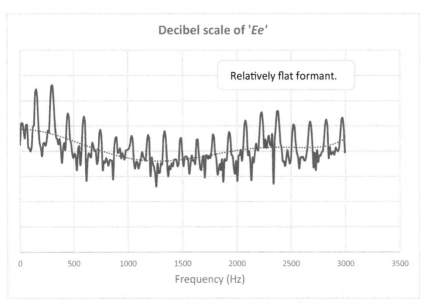

**Figure 3.41.** High-pitch (male) sound of '*ee*'.

### 3.7.3 Human voice when ill

Related to the human voice sound mechanism, there is another interesting observation. A person's voice was recorded when he was healthy and ill. Figures 3.42–3.45 show normalized spectra of the person's healthy and sick '*ah*'

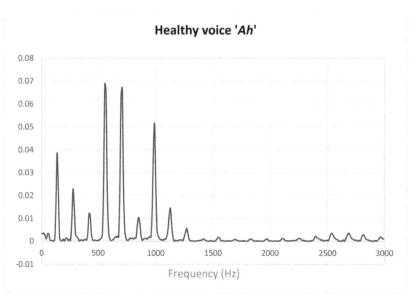

**Figure 3.42.** Healthy male voice sound of '*ah*' (normalized).

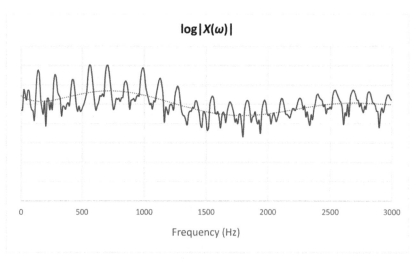

**Figure 3.43.** Normal male voice sound of '*ah*'(normalized).

voices. While the generating sound frequencies do not change appreciably, the large change in the suppression of intensities accounts for the large damping constant due to the swollen tissues of the oral cavity.

It is the ultimate challenge for voice recognition to identify a person's voice under various physical conditions in which the peak frequencies and the formant are both changed.

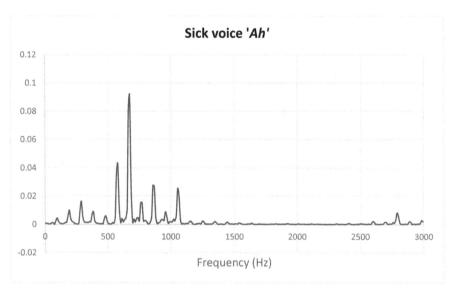

**Figure 3.44.** Sick voice sound of 'ah' (normalized).

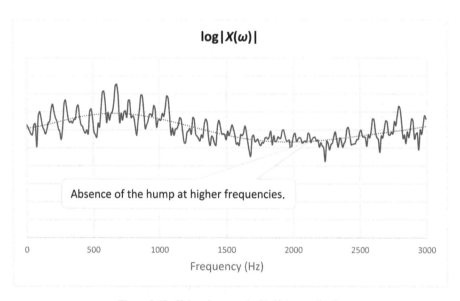

**Figure 3.45.** Sick voice sound of 'ah' (normalized).

### 3.7.4 Musical instruments

**Harmonica**: figure 3.46 shows a composite sound of a harmonica blowing multiple slots at once. The sampling frequency is 16 kHz, and the number of data points is 2048. Therefore, the frequency resolution is (16 000)/2048 ≈ 7.813 Hz.

Figure 3.47 is the Fourier transform spectrum with application of the DC offset and the Hanning windowing function.

**Figure 3.46.** Harmonica sound from multiple slots.

**Figure 3.47.** Harmonica sound spectrum (2048-point fast Fourier transform (FFT)).

**Glass cup:** figure 3.48 is the sound from a drinking glass hit with a wooden stick. Similar to percussion instruments, this sound can be considered to be an impulse response (figure 3.49).

**Remark**: those who are interested in investigating musical instrument spectra can utilize numerous sound clips available from the internet [3].

### 3.7.5 Cepstrum analysis

Rather than the intuitive polynomial fitting to the logarithmic amplitude, there is a better analytical method for characterizing the formant. Similar to the Fourier transform, we can attempt to analyze the formant by performing the inverse Fourier

**Figure 3.48.** Impulse sound from a drinking glass.

**Figure 3.49.** Energy spectrum.

transform of the logarithmic spectrum. This idea is the foundation of the cepstrum discussed in section 2.5. With the cepstrum, the effects of the vocal cords and the vocal tract can be differentiated.

Of the logarithmic power spectrum of figures 3.33 and 3.35, the broad spectral envelope characterizes the vocal tract and the peaks correspond to the frequencies by the vibrations of the vocal cords. By carrying out the *inverse* Fourier transform of the logarithmic amplitude, $\log|Y(\omega)|$, the cepstrum of the human voice is obtained as shown in figure 3.50. The cepstrum shown here is the real part of the inverse Fourier transform. The lower quefrency region is the cepstrum of the vocal tract, $c_g(m)$, and

**Figure 3.50.** Cepstrum of logarithmic spectrum log| $Y(\omega)$|.

the higher quefrency region is the cepstrum of the vocal cord vibration, $c_x(m)$. Detailed cepstrum analysis is beyond the scope of this book, and readers should refer to advanced articles [6].

## References

[1] For example, *Moo0 Voice Recorder* (http://download.cnet.com/Moo0-Audio-Recorder/3000-2170_4–10864751.html)

[2] For example, *AVSAudioEdiotor* (http://download.cnet.com/AVS-Audio-Editor/3000-2170_4–10406673.html)

[3] For example, *WaVCsvWav* (file:///L:/AcousticAnalysis/Sound%20Analysis%20Program/WavCsVWav/document/usage.html)

[4] *Understanding FFTs and Windowing* National Instruments (http://ni.com/white-paper/4844/en/)

[5] Gasior M and Gonzalez J L *Improving FFT Frequency Measurement Resolution by Parabolic and Gaussian Interpolation* CERN-AB Division, AB-Note-2004-021 BDI 2005 (http://cds.cern.ch/record/720344/files/ab-note-2004-021.pdf)

[6] L9: *Cepstral Analysis* Texas A&M University (https://people.engr.tamu.edu›rgutier›web_-courses›csce630_f14›l9.pdf)

# Part 2

Advanced theories and topics on Fourier transfrom

**IOP** Publishing

Fourier Transform and Its Applications Using Microsoft EXCEL®
(Second Edition)

Shinil Cho

# Chapter 4

# Applications of Fourier transform in physics

There is a mathematical method for finding frequency components and their amplitudes, or spectra from periodic patterns. It is called Fourier transform (FT) spectral analysis. In this chapter, how we apply Fourier transform to the following subjects will be explored, emphasizing theoretical discussions:

1. electronics circuits and telecommunication;
2. spectroscopy—magnetic resonance, and IR-spectroscopy;
3. optics—one- and two-dimensional diffractions, lens, hologram, and x-ray diffraction;
4. stochastic process—central limit theorem, Brownian motion, fluctuation–dissipation theorem, and motional narrowing;
5. spectral method for solving diffusion equation and Green's function for wave equation; and
6. quantum mechanics—wave packet, group velocity, imaginary time Schrödinger equation, Heisenberg's uncertainty principle, and the Born approximation of scattering of a particle of potential.

## 4.1 Electronic circuits

Electronic circuits provide good examples of the transfer function $G(\omega)$ of the linear black box theory discussed in chapter 2. Here, we analyze $CR$, $RC$, $LR$, and RLC-series circuits in the time and the frequency domains [1].

### 4.1.1 *CR*-series circuit

Figure 4.1 is a circuit diagram of a $CR$-series circuit. For the input signal of $V_i = x(t)$, the output signal of $V_o = y(t)$ is the voltage difference across the resistor $R$. The equation between the input and the output for this circuit is given by

**Figure 4.1.** CR-series circuit.

$$V_i = \frac{1}{RC} \int V_o dt + V_o \tag{4.1}$$

or

$$x(t) = \frac{1}{RC} \int y(t)dt + y(t). \tag{4.2}$$

Thus, we obtain the following differential equation:

$$\frac{dx(t)}{dt} = \frac{1}{RC}y(t) + \frac{dy(t)}{dt}. \tag{4.3}$$

Using the differentiation property of the Fourier transform (equation (2.21)), we obtain from equation (4.3)

$$i\omega X(\omega) = \frac{1}{RC}Y(\omega) + i\omega Y(\omega) \tag{4.4}$$

where $x(t) = \frac{1}{2\pi} \int_{-\infty}^{+\infty} X(\omega)e^{i\omega t}d\omega$ and $y(t) = \frac{1}{2\pi} \int_{-\infty}^{+\infty} Y(\omega)e^{i\omega t}d\omega$.
From equation (4.4), the output signal $Y(\omega)$ is given by

$$Y(\omega) = \frac{i\omega RC}{1 + i\omega RC}X(\omega). \tag{4.5}$$

Therefore, the transfer function $G(\omega) = Y(\omega)/X(\omega)$ for the CR-series circuit is given by

$$G(\omega) = \frac{i\omega RC}{1 + i\omega RC}. \tag{4.6}$$

Using the transfer function $G(\omega)$, the output signal across the resistor $R$ in the time domain can be computed by the inverse Fourier transform,

$$y(t) = \int_{-\infty}^{+\infty} G(\omega)X(\omega)e^{-i\omega t}d\omega. \tag{4.7}$$

Let us investigate the output signal when the input signal $x(t)$ is a step function. The Fourier transform of the step function shown below is given by equation (2.20). The output signal $y(t)$ is thus given by

$$y(t) = \frac{1}{2\pi} \int_{-\infty}^{+\infty} G(\omega)U(\omega)e^{-i\omega t}d\omega$$

$$= \int_{-\infty}^{+\infty} \frac{1}{2\pi i\omega} \frac{i\omega RC}{1 + i\omega RC}e^{-i\omega t}dt + \frac{1}{2}\left[\frac{i\omega RC}{1 + i\omega RC}\right]_{\omega=0} \quad (4.8)$$

$$= \frac{1}{2\pi i} \int_{-\infty}^{+\infty} \frac{e^{-i\omega t}}{\omega - i(RC)^{-1}}d\omega = e^{-\frac{t}{RC}}$$

by applying the Cauchy theorem (appendix A2).

Since the voltage across the resistor $R$ is $V_0 = y(t)$ and the input voltage $V_i$ for $t > 0$ is 1, the voltage across the capacitor $C$ in the time domain is given by

$$V_C = V_i - V_R = 1 - e^{-\frac{t}{RC}}. \quad (4.9)$$

This is a well-known result from the $CR$-series circuit, and $\tau = RC$ is the time constant of the circuit. The advantage of the linear response analysis is that the output function can be calculated for an arbitrary input function systematically with the transfer function. Figure 4.2 shows $V_i$ and $V_c$ when the time constant is 5 s.

The transfer function $G(\omega)$ of a circuit is the characteristics of frequency response of the circuit. The $CR$-series circuit can be regarded as a high-pass filter in the frequency domain. The transfer function (4.6) can be modified to

$$G(\omega) = \frac{i\omega/\omega_0}{1 + i(\omega/\omega_0)} = \frac{1}{1 - i(\omega_0/\omega)} \quad \text{where } \omega_0 = 1/RC.$$

**Figure 4.2.** Stepwise input voltage of and voltage across capacitor.

Thus,

$$|G(\omega)| = \frac{(\omega/\omega_0)}{\sqrt{1 + (\omega/\omega_0)^2}}. \tag{4.10}$$

Figure 4.3 shows $|G(\omega)|$ as a function of $\omega/\omega_0$, which characterizes a high-pass filter.

### 4.1.2 *RC*-series circuit

Figure 4.4 shows a *RC*-series circuit. In this circuit, the output is the voltage difference across the capacitor *C*, and the equation between the input and the output is given by

$$V_i = CR\frac{dV_0}{dt} + V_0. \tag{4.11}$$

The Fourier transform of equation (4.11) leads to

$$X(\omega) = (1 + i\omega RC)Y(\omega) \tag{4.12}$$

and the transfer function $G(\omega)$ is given by the following equation:

$$G(\omega) = \frac{1}{1 + i(\omega/\omega_0)} \text{ and } |G(\omega)| = \frac{1}{\sqrt{1 + (\omega/\omega_0)^2}} \text{ where } \omega_0 = 1/RC. \tag{4.13}$$

In the time domain, the response to the step function is given by

**Figure 4.3.** $|G(\omega)|$ of the *CR*-series circuit.

**Figure 4.4.** *RC*-series circuit.

**Figure 4.5.** Stepwise input voltage and voltage across capacitor.

$$y(t) = \frac{1}{2\pi} \int_{-\infty}^{+\infty} G(\omega) \left[ \pi\delta(\omega - \frac{i}{\omega}) \right] e^{-i\omega t} d\omega$$

$$= \int_{-\infty}^{+\infty} \frac{1}{2\pi i} \frac{1}{1 + i\left(\frac{\omega}{\omega_0}\right)} e^{-i\omega t} dt + \frac{1}{2} \left[ \frac{1}{1 + i\left(\frac{\omega}{\omega_0}\right)} \right]_{\omega=0} = 1 - e^{-\frac{t}{RC}}. \qquad (4.14)$$

The output voltage drop across the resistor $R$ is thus given by $V_R = V_i - V_c = e^{-\frac{t}{RC}}$ in the time domain, and the time constant is $\tau = RC$. Figure 4.5 shows $V_i$ and $V_C$ when the time constant $RC$ is 5 s.

In the frequency domain, the transfer function of the $RC$-series circuit exhibits a characteristic of low-pass filter. Figure 4.6 shows $|G(\omega)|$ as a function of $\omega/\omega_0$.

**Figure 4.6.** $|G(\omega)|$ of the $RC$-series circuit.

**Figure 4.7.** $LR$-series circuit.

### 4.1.3 *LR*-series circuit

Figure 4.7 is a circuit diagram of a *LR*-series circuit. For the input signal $V_i = x(t)$, the output signal $V_o = y(t)$ is the voltage difference across the resistor $R$.

The equation between the input and the output is given by

$$V_i = \frac{L}{R}\frac{dV_o}{dt} + V_o \text{ and } V_o = I \times R$$

or

$$x(t) = \frac{L}{R}\frac{dy(t)}{dt} + y(t). \tag{4.15}$$

The Fourier transform of the above equation is

$$X(\omega) = i\omega\frac{L}{R}Y(\omega) + Y(\omega), \text{ and hence } Y(\omega) = \frac{1}{1 + i\omega(L/R)}X(\omega). \tag{4.16}$$

and the transfer function of the *LR*-series circuit is

$$G(\omega) = \left(1 + i\omega\frac{L}{R}\right)^{-1} = \left(1 + i\frac{\omega}{\omega_0}\right)^{-1} \quad \text{where } \omega_0 = 1/(L/R). \quad (4.17)$$

The output is given by equation (4.16). If the input is the step function, the output voltage across the resistor *R* becomes

$$y(t) = \frac{1}{2\pi}\int_{-\infty}^{+\infty} G(\omega)X(\omega)e^{-i\omega t}dt = \int_{-\infty}^{+\infty}\frac{1}{2\pi i}\frac{1}{\omega}\frac{1}{1+i\omega\frac{L}{R}}e^{-i\omega t}dt + \frac{1}{2}\left[\frac{1}{1+i\omega\frac{L}{R}}\right]_{\omega=0}$$

$$= \frac{1}{2\pi i}\int_{-\infty}^{+\infty}\left[\frac{1}{\omega} - \frac{1}{\omega + i(L/R)}\right]e^{-i\omega t}dt + \frac{1}{2} \quad (4.18)$$

$$= \frac{1}{2} - e^{-\frac{t}{(L/R)}} + \frac{1}{2} = 1 - e^{-\frac{t}{(L/R)}}.$$

where $\tau = L/R$ is the time constant of the circuit. Since the voltage difference across the resistor *R* is given by equation (4.18), the voltage across the inductor *L* is given by

$$V_L = V_i - V_R = 1 - y(t) = e^{-\frac{t}{(L/R)}}.$$

These results are also a well-known for the response of the *LR* circuit to the input of the step function. The response $V_L$ to a step-wise input of this circuit is the same as the CR-circuit as shown in figure 4.6 when the time constant $L/R$ is 5 s.

Comparing the transfer function of the *LR*-series circuit with that of the *RC*-series circuit, both are the same form, and thus, the *LR*-series circuit is also characterized as a low-pass filter in the frequency domain.

### 4.1.4 *RLC*-series circuit

For an *RLC*-series circuit shown in figure 4.8, we apply slightly different approach from the *RC* and the *RL* circuits. The differential equation of the LRC-series circuit is given by

$$L\frac{d^2q}{dt^2} + R\frac{dq}{dt} + \frac{q}{C} = V_i(t) \quad (4.19)$$

where $q(t)$ is the charge on the capacitor *C* and $V_i(t)$ is the input voltage in the time domain.

Equation (4.19) is the same as the differential equation (2.57) for the forced harmonic oscillator with variable replacements:

$$M \to L, \ M\omega_0^2 \to 1/C \text{ or } \omega_0^2 = 1/LC, \text{ and } M\Gamma \to R \text{ or } \Gamma = R/L. \quad (4.20)$$

Therefore, we can use equations (2.57)–(2.60) to obtain the following equations:

$$Q(\omega) = G(\omega)V_i(\omega) \quad (4.21)$$

where $Q(\omega)$ and $V_i(\omega)$ are Fourier transform of $q(t)$ and $V_i(t)$, and $G(\omega)$ is the frequency transfer function given by

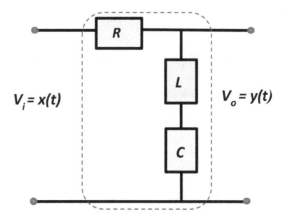

**Figure 4.8.** *RLC*-series circuit.

$$G(\omega) = \frac{1}{i\omega Z(\omega)} \tag{4.22}$$

where

$$Z(\omega) = R + i\left(\omega L - \frac{1}{\omega C}\right) \tag{4.23}$$

is the impedance of the circuit.

Because the electrical current is given by $I(t) = dq(t)/dt$, the Fourier transform of the current is given by $I(\omega) = i\omega Q(\omega) = V_i(\omega)/Z(\omega)$ from equations (4.21), (4.22), and (2.19). The voltage $V_{LC}(t)$ across $L$ and $C$ is given by $V_{LC}(t) = V_i(t) - R\cdot I(t)$, and its Fourier transform is

$$V_{LC}(\omega) = V_i(\omega) - RI(\omega) = [1 - R/Z(\omega)]V_i(\omega). \tag{4.24}$$

From equations (4.23) and (4.24), $|Z(\omega)|$ becomes its minimum value of $R$, and $V_{LC}(\omega) = 0$ and $I(\omega) = V_i(\omega)/R$ when $\omega^2 = \omega_0{}^2 = 1/LC$. Therefore, the *RLC*-series circuit is a band-stop filter. Figure 4.9 shows $|Z(\omega/\omega_0)| = [R^2 + (\omega L)^2(1 - (\omega_0/\omega)^2)^2]^{(1/2)}$ with $R = 0$, $L = 1$, and $\omega_0 = 1$.

From the admittance given by equations (2.73), the admittance of the *RLC* circuit is given by

$$\chi(\omega) = \frac{1}{L}\frac{1}{(\omega^2 - \omega_0{}^2 - i\Gamma\omega)} \quad \text{where } \omega_0{}^2 = 1/LC \text{ and } \Gamma = R/L. \tag{4.25}$$

### 4.1.5 Note on filtering

Filters in electronic circuits eliminate or select a specific frequency interval. Such filtering functions can be also performed with Fourier transform once a signal in the time domain is stored and converted into the frequency domain. Sometimes the software filter can be implanted what we cannot build by hardware. For example, the ideal band-pass filter has a sharp cut in the frequency domain:

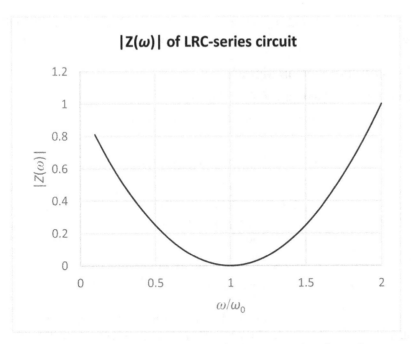

**Figure 4.9.** $|Z(\omega/\omega_0)|$ of the *LRC*-series circuit ($R=0$, $L = 1$, and $\omega_0 = 1$).

$$G(\omega) = \text{constant} = D \text{ if}|\omega| \leqslant \omega_c; \text{ and } 0 \text{ otherwise.} \qquad (4.26)$$

From equation (2.56) of the linear response theory, the output signal in the time domain is given by

$$y(t) = \int_{-\infty}^{\infty} x(t - \tau)g(\tau)d\tau. \qquad (4.27)$$

Because the impulse response $g(\tau)$ is the inverse Fourier transform of $G(\omega)$, we obtain

$$g(\tau) = \frac{1}{2\pi} \int_{-\infty}^{\infty} G(\omega)e^{i\omega\tau}d\omega = \frac{1}{2\pi} \int_{-\omega_c}^{\omega_c} De^{i\omega\tau}d\omega = \frac{\omega_c D}{\pi} \frac{\sin(\omega_c\tau)}{\omega_c\tau}. \qquad (4.28)$$

In this case, the impulse response $g(\tau)$ is a sinc-function (4.28), which does not vanish even if $\tau<0$. In other words, the ideal filter does not follow causality, and therefore, it is physically impossible to materialize such an ideal filter using physical circuit elements. Nevertheless, it is possible to implement the equivalent digital filtering capability on a digitally recorded signal. For digital filtering, equation (4.27) is modified to a weighted moving average:

$$y_j = \sum_{j=-N}^{N} g_k x_{j-k} \text{ where } j = 1, 2,..., m. \qquad (4.29)$$

where $\{x_j = x(j\Delta t)$ and $g(j\Delta t); j = 1, 2, ..., m\}$ are digitized sequences of $x(t)$ and $g(t)$ at discrete time $t = j\Delta t$. For a detailed discussion of how to design digital filters, refer to [2].

## 4.2 Telecommunication signals

### 4.2.1 Amplitude modulation and demodulation

An amplitude modulation (AM) signal can be described by the following theoretical equation

$$x_{AM}(t) = A_0[1 + m \sin(2\pi f_m t)] \sin(2\pi f_c t) \tag{4.30}$$

where $A_0$, $m$, $f_m$, and $f_c$ are the amplitude, the degree of modulation, the signal frequency, and the carrier frequency. The signal frequency is usually smaller than the carrier frequency, i.e. $f_m < f_c$. and Figure 4.10 shows a simulated signal of $A_0 = 1$, $m = 0.4$, $f_m = 10$ Hz, and $f_c = 100$ Hz.

Equation (4.30) is equivalent to

$$x_{AM}(t) = A_0 \sin(2\pi f_c t) + \frac{1}{2}A_0 m \cos[2\pi(f_c - f_m)t] - \frac{1}{2}A_0 m \cos[2\pi(f_c + f_m)t]. \tag{4.31}$$

From equation (4.31), we can predict that there should be three frequency components of $f_c$ and $f_c \pm f_m$. Figure 4.11 shows the spectrum of the AM signal generated by EXCEL where we changed the initial phase of the above signal by $\pi/2$ for the fast Fourier transform (FFT). The spectrum remains the same because of the time invariance property (section 2.1.2). In this spectrum, the time step is 0.25 ms and there are 4096 data points. The frequency resolution is thus $1/(0.25 \times 10^{-3})$ $(4096) = 0.9766$ Hz.

**Figure 4.10.** Amplitude modulation signal.

**AM signal spectrum from FFT**

$f_c$ = 100 Hz
$f_m$ = 10 Hz
$m$ = 0.4

$f_c \pm f_m$ = 90, 110 Hz

Frequency (Hz)

**Figure 4.11.** AM signal spectrum by Fourier transform.

The theoretical Fourier transform of the AM signal given by equation (4.30) can be calculated by

$$X_{AM}(f) = \frac{A_0}{2}[\delta(2\pi(f - f_c)) + \delta(2\pi(f + f_c))]$$
$$+ \frac{m}{2}[\delta(2\pi(f - f_c - f_m)) + \delta(2\pi(f + f_c + f_m)) \quad (4.32)$$
$$+ \delta(2\pi(f - f_c + f_m)) + \delta(2\pi(f + f_c - f_m))].$$

Dropping the negative frequency components, $(-f_c+f_m)$, we obtain the relevant components to the spectrum.

$$X_{AM}(f) = \frac{A_0}{2}\delta(2\pi(f - f_c)) + \frac{mA_0}{2}[\delta(2\pi(f - f_c - f_m)) + \delta(2\pi(f - f_c + f_m))]. \quad (4.33)$$

Figure 4.12 shows the theoretical spectrum given by equation (4.33) where the delta function is defined as $\delta(f - f_c) = 1$ if $f = f_c$, and 0 otherwise for displaying purposes.

**Demodulation of AM signal:** an AM signal can be demodulated by mixing (or multiplying) the sine-function of the carrier frequency. Let us start with a general case where a signal $M(t)$ is modulated to be $x_{AM}(t) = [1 + M(t)]\sin(\omega_c t)$.

$$[x_{AM}(t)]\sin(\omega_c t) = [1 + M(t)]\sin^2(\omega_c t)$$
$$= (1/2)[1 + M(t)] - (1/2)[1 + M(t)]\cos(2\omega_c t). \quad (4.34)$$

Because the frequency band of $M(t)$ is much lower than the carrier frequency, the second term of equation (4.34) can be low-pass filtered out to obtain $(1/2)[1 + M(t)]$. Then, $M(t)$ can be reproduced by cutting the DC component. Once the DC component of the low-pass filtered signal is eliminated, $M(t)$ can be reproduced. The early demodulation method is the square law detection using a diode where

**Theoeretical AM spetcrum**

$f_c = 100$ Hz
$f_m = 10$ Hz
$m = 0.4$

$f_c \pm f_m = 90$
and 110 Hz

The theoretical amplitude of the central peak to the side band is set to be 50 (=1/0.2) because m = 0.4 in this example.

Frequency (Hz)

**Figure 4.12.** The theoretical spectrum with a frequency resolution of 1 Hz.

$[x_{AM}(t)]^2 = [1 + M(t)]^2 \sin^2(\omega_c t)$ is created and the $\cos(2\omega_c t)$-term is eliminated by a low-pass filter. However, due to the squared amplitude, there will be the second harmonics of $M(t)$ as the signal distortion.

### 4.2.2 Frequency modulation and demodulation

A frequency modulation (FM) signal can be represented by the following theoretical equation

$$x_{FM}(t) = A_0 \sin[2\pi f_c t + m \sin(2\pi f_m t)] \tag{4.35}$$

where $A$, $m$, $\delta$, and $f_c$ are the amplitude, the degree of modulation, the signal frequency, and the carrier frequency. Figure 4.13 shows the simulated signal of $A_0 = 1$, $m = 0.8$, $f_m = 50$ Hz, and $f_c = 100$ Hz.

Figure 4.14 shows the spectrum of the FM signal generated by EXCEL, which is rather complicated, around the carrier frequency $f_c$. The time step = 0.025 ms and 4096 data points. The frequency resolution is thus 9.766 Hz.

**Remark**: The frequency distribution of the FM signal can be expressed with the Bessel function [3]:

$$
\begin{aligned}
x_{FM}(t) = A_0[ &\sum_{n=-\infty}^{+\infty} J_n(m)\sin\{2\pi(f_c + nf_m)t\}] \\
= A_0[ &J_0(m)\sin(2\pi f_c t) + J_1(m)\{\sin(2\pi(f_c + f_m)t) - \sin(2\pi(f_c - f_m)t\} \\
&+ J_2(m)\{\sin(2\pi(f_c + 2f_m)t) + \sin(2\pi(f_c - 2f_m)t\} \\
&+ J_3(m)\{\sin(2\pi(f_c + 3f_m)t) - \sin(2\pi(f_c - 3f_m)t\} \\
&+ \cdots].
\end{aligned} \tag{4.36}
$$

**Figure 4.13.** Frequency modulation signal.

**Figure 4.14.** Fourier transform of FM signal.

Calculation of the Fourier transform gives a discrete FM spectrum with magnitude coefficients as a function of $m$, as shown in the equation below.

$$X_{\text{FM}}(f) = \int_{-\infty}^{+\infty} x_{\text{FM}}(t)e^{-i2\pi ft}dt = \frac{A_0}{2}\sum_{n=-\infty}^{n=+\infty} J_n(m)[\delta(f - f_c - nf_m) + \delta(f + f_c + nf_m)]$$

$$= \frac{A_0}{2}[J_0(m)\{\delta(f - f_c) + \delta(f + f_c)\} + J_1(m)\{\delta(f - f_c - f_m) + \delta(f - f_c + f_m)\}$$

$$+ J_2(m)\{\delta(f - f_c - 2f_m) + \delta(f - f_c + 2f_m)\}$$

$$+ J_3(m)\{\delta(f - f_c - 3f_m) + \delta(f - f_c + 3f_m)\} + \cdots],$$

$$+ \cdots], \tag{4.37}$$

where $\delta(f - f_c - nf_m)$ and $\delta(f - f_c + nf_m)$ are the delta-functions. If $f_c = 100$ Hz and $f_m = 50$ Hz, the spectrum of the FM signal has the frequency components of $\pm f_c = \pm 100$ Hz, $(f_c \pm f_m) = 150$ Hz or 50 Hz, $(f_c \pm 2f_m) = 0$, 200 Hz, $(f_c \pm 3f_m) = -50$ Hz (not possible), 250 Hz, and so forth.

Figure 4.29 shows the FM spectrum based on equation (4.37). The vertical scale is change for comparison with the theoretical FM spectrum shown in figure 4.15.

Since EXCEL has built-in Bessel functions (BESSELI, BESSELJ, BESSELK, and BESSELY), the theoretical spectrum can be generated with EXCEL. Figure 4.30 shows these frequency components with a comparable frequency resolution to the spectrum by FFT.

**Demodulation of FM signal:** There are several types of FM demodulation methods. Their basic principle is to convert the frequency variations of a FM signal from the carrier frequency $f_c$ to the amplitude variations and then use an AM demodulation. A simple FM demodulation can be performed by taking a time average of a train of zero-crossing pulses of FM signal. Each pulse of the pulse train

**Figure 4.15.** Theoretical spectrum of FM signal.

is located at a zero-crossing of the FM signal, and the pulse train is called pulse-repetition-rate modulated signal. A low-pass filter is use to average (integral) the pulse train to produce an AM signal. For detailed descriptions of FM demodulations, refer to electronics or telecommunication books [4].

## 4.3 Spectroscopy

### 4.3.1 Nuclear magnetic resonance spectroscopy

Imagine that an isolated nuclear spin, e.g. a proton, is placed in an external magnetic field $B$. Its energy levels are given by $U = -\vec{\mu} \cdot \vec{B}$ where $\mu$ is the magnetic moment of the nuclear spin. For a proton, the magnetic moment is $2.8\mu_n$ where $\mu_n$ is the nuclear magneton. Because a proton has nuclear spin 1/2, it exhibits two energy levels split with the application of a magnetic field, as shown in figure 4.16. The proton is usually at the lower energy state, but it can be excited to the higher energy state if the energy $\Delta U = 2\mu B_0$ to the proton is supplied externally to cause the transition. The excitation energy can be in the form of an electromagnetic wave of frequency $f$ to cause the transition

$$\Delta U = 2\mu B_0 = \hbar\omega_0 \tag{4.38}$$

where $\hbar$ is the Planck constant. For example, when $B_0 = 1$ Tesla, the electromagnetic frequency for a proton is 42.65 MHz. Therefore, a proton in a magnetic field of 1 Tesla and irradiated with an electromagnetic wave of 42.65 MHz will undergo the energy transition.

This is called nuclear (spin) magnetic resonance (NMR), and the electromagnetic frequency is called the Larmor frequency.

Protons in a substance in an external magnetic field will also experience local magnetic fields—denoted as $\Delta B$—due to the surrounding environmental onset to the uniform external field. Thus, the net magnetic field experienced by a proton is $\Delta B + B_0$. The distribution of spin energy states reflects the surrounding structure, which is what we call the 'spin–spin interaction.' In other words, the resonance frequencies are slightly different (modulated) from the Larmor frequency, becoming $\Delta\omega + \omega_0$.

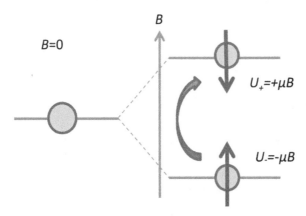

**Figure 4.16.** Magnetic potential energy.

**Figure 4.17.** Selective excitation of pendulum system.

By sweeping the frequency of the irradiating electromagnetic wave around the Larmor frequency, energy absorptions will be observed at each of the resonance frequencies, $\Delta\omega + \omega_0$, successively. This is called the continuous wave nuclear magnetic resonance (CW-NMR). This situation is analogous to observed resonance frequencies of a set of simple pendulums shown in figure 4.17. In this photograph, only the middle pendulum is selectively excited to resonate the frequency of the external driving force to that of the middle pendulum. If the frequency of the external driving force is gradually changed, we can observe resonance oscillations one after another. However, this approach has a much longer measurement time.

### 4.3.2 Pulse NMR

There is another way to excite *all* resonances of the pendulum system simultaneously. By applying an intense force in a pulse, all of the pendulums can be oscillated simultaneously. This situation is analogous to the pulse NMR. In a pulse NMR, an intense electric pulse of the Larmor frequency is applied to the target nucleus, e.g. the proton, to cause all possible transitions at once. This is because an electromagnetic pulse, called an RF pulse as it is in the radio frequency range, covers a wide frequency range around the Larmor frequency. How can a single RF pulse contain a frequency spectrum around the Larmor frequency? The frequency spectrum of the delta function $\delta(t)$ ranges the entire frequency, and thus it is an ideal pulse. In experimental apparatuses, a power transmitter sends an intense RF pulse to an NMR sample. The RF pulse will generate a magnetic field from a

transmission coil for a short time, exciting the nuclear spin to higher energy states (or flipping nuclear spins).

The energy spectrum of an RF pulse is shown in figure 4.18. This square RF pulse has Larmor frequency 10 MHz and pulse width 0.25 μs.

Figure 4.19 shows the spectrum of the square RF pulse. The RF pulse has an energy spectrum with a frequency distribution of around 10 MHz. Due to the finite

**Figure 4.18.** RF pulse.

**Figure 4.19.** Energy spectrum of an RF pulse.

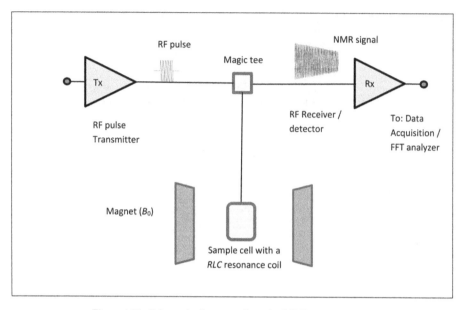

**Figure 4.20.** Schematic diagram of a pulse NMR spectrometer.

size of the RF pulse, the energy spectrum is not quite symmetric and there is a windowing effect around the Larmor frequency. If the frequency distribution is broad enough, we can excite or pump all nuclear spins to higher energy states.

Once the proton is pumped to a higher energy level, it will emit photons of the frequencies returning to the lower energy levels of the thermal equilibrium state. The emitting electromagnetic signal is composed of various transition frequencies of the fine structure of energy levels due to the spin–spin interaction. The signal frequencies are onset to the Larmor frequency. This signal, called the free induction decay (FID), is a response to the RF pulse of the Larmor frequency, and is detected by a receiver coil. The FID signal is an AM signal with the Larmor frequency in the time domain, and the signal amplitude reflects the local magnetic fields due to the spin–spin interaction. FID is Fourier transformed to find the NMR frequency spectrum [5]. Figure 4.20 depicts a schematic diagram of a pulse NMR spectrometer.

Figures 4.21 and 4.22 show an FID and its Fourier transformed spectrum from a water sample. It should be pointed out that the sound signal generated by hitting a drinking glass (figure 3.21) exhibits a similar time dependence and spectrum due to *impulse* response (figure 3.22).

### 4.3.3 Fourier transform IR (FT–IR) spectroscopy

FT–IR spectrum analysis is particularly useful for analyzing the infrared absorption of gases where the spectra is extremely complicated [6]. The method acquires interferograms using the Michelson–Morley interferometer. The acquired interferometer is then Fourier transformed to obtain spectra.

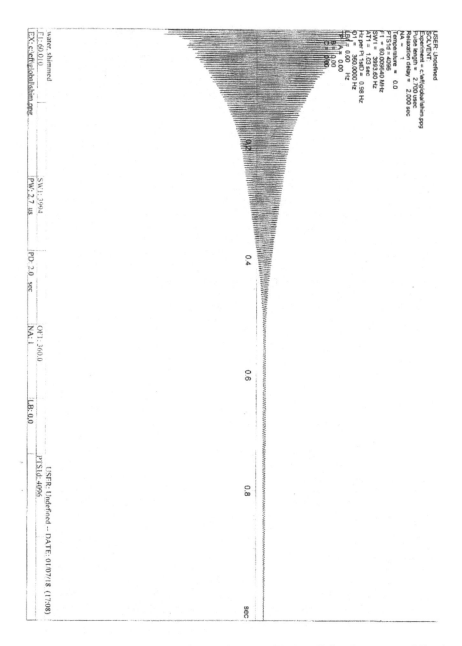

**Figure 4.21.** Free induction decay (FID) of water. Courtesy of Dr Don Fujito, Department of Chemistry, La Roche University.

In figure 4.23, if the light is not monochromatic but has a spectral composition given by the function $G(\omega)$, the intensity at the sample cell changes in a manner that depends upon the particular spectrum. Recording the intensity as a function of path difference can deduce the power spectrum $G(\omega)$. This method for obtaining a spectrum is called Fourier transform IR (FT–IR) spectroscopy.

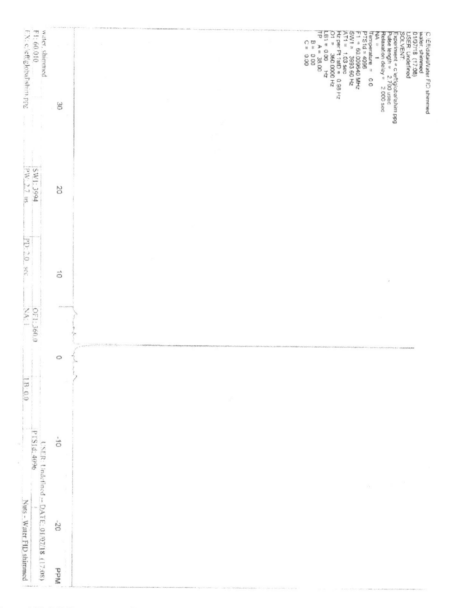

**Figure 4.22.** NMR spectrum of water. Courtesy of Dr Don Fujito, Department of Chemistry, La Roche University.

Since the FT–IR commonly uses the wave number $k$ rather than the angular frequency $\omega = ck$, the power spectrum $G(k)$ is a function of the wave number. The intensity for non-monochromatic light at the sample cell will be given by a summation over the complete spectrum:

$$I(x) = \int_0^\infty (1 + \cos kx)G(k)dk = \int_0^\infty G(k)dk + \frac{1}{2}\int_{-\infty}^\infty G(k)e^{ikx}dk \qquad (4.39)$$

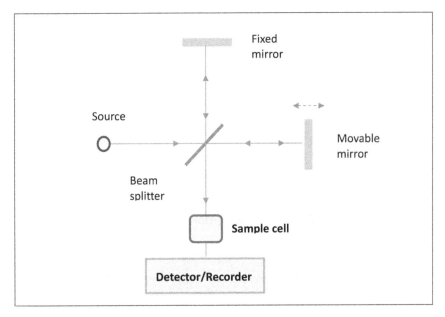

**Figure 4.23.** Schematic diagram of a FT–IR spectrometer.

where the second term $\int_0^\infty G(k)dk$ is the intensity for zero path difference. Therefore,

$$W(x) \equiv 2I(x) - \int_0^\infty G(k)dk = \int_{-\infty}^{+\infty} G(k)e^{ikx}dk. \tag{4.40}$$

Measuring the intensity $W(x)$—called the interferogram—as a function of the path difference $x$, the power spectrum can be calculated from the Fourier transform of $W$ $(x)$:

$$G(k) = \frac{1}{2\pi} \int_{-\infty}^{+\infty} W(x)e^{-ikx}dk. \tag{4.41}$$

In an actual absorption FT–IR spectrum, these molecular vibrations cause IR absorptions, and the absorption spectrum will be $1 - (I/I_0)$, where $I_0$ is the total input intensity and $I$ is the output intensity at a given frequency. The total input intensity is calculated by intergradation over the observed frequency range.

The Fourier transform of a harmonica sound exhibits an analogous multiple sound 'interferogram' as shown in figure 4.24, and an 'absorption' spectrum of the harmonica sound generated by EXCEL's FFT is shown in figure 4.25.

## 4.4 Optics

Wave optics, and diffraction in particular, is an ideal subject to study the Fourier transform [7]. Light is an electromagnetic wave, and it has two vector quantities: the electric field and the magnetic field. However, in the optical diffraction, when an aperture size is larger than the wavelength of light, if we observe the diffraction

**Figure 4.24.** Sound interferogram of a multiple sound from a harmonica.

## Simulated Absorption Spectrum

The simulated absorption is $1-(I/I_0)$ where $I_o$ is the total signal enengy.

**Figure 4.25.** Simulated absorption spectrum.

sufficiently far away from the aperture, we can consider that the electric field and the magnetic field independently vary. Furthermore, we only need to consider the field magnitude, i.e. the scalar field approximation is applicable, and the 'light wave' can be used to describe the diffraction.

### 4.4.1 One-dimensional diffraction

First, we consider a one-dimensional case. In figure 4.26, the amplitude of light observed at point $P$ on the screen is superposing or integrating the plane wave $e^{ikr}$ over the aperture. Alternatively, we can interpret this equation as the Fourier transform.

Points $O$ and $Q$ are the origin and an arbitrary point of the one-dimensional slit, respectively. The position of point $Q$ is expressed by the coordinate $x$ along the

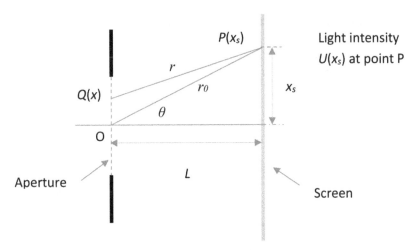

**Figure 4.26.** Single slit diffraction.

aperture from the origin $O$. The observing point $P$ is on the screen located behind the aperture. The distance from the aperture to the screen, $L$, is usually very large for the Fraunhofer diffraction although the figure does not accurately represent the distance. Taking the $x_s$-axis along the screen, aside from the constant factor, the light intensity originates from the element of the segment $dx$. In this case, the one-dimensional diffraction has the more general formula, $U(x_s)$ at the observing point $P$ ($x_s$)—where $x_s$ is the distance from the center of the screen to the point $P$—is given by the following equation:

$$U(x_s) = \int e^{ik(r_0+\delta x)}dx = e^{ikr_0}\int e^{ik\delta x}dx \Rightarrow \int e^{ik\delta x}dx \qquad (4.42)$$

where $r$ is the distance between points $Q$ and $P$, $r_0$ is the distance between points $O$ and $P$, and $\delta x$ is the path difference between $r$ and $r_0$. For the Fraunhofer diffraction, we can apply the small angle approximation, i.e.

$$\delta x = x\left(\frac{x_s}{L}\right) = x\sin\theta. \qquad (4.43)$$

For an aperture with a complicated geometry, we introduce a function, $g(x)$—the aperture function. This function is defined such that $g(x)dx$ is the amplitude of the diffraction light

$$U(x_s) = \int g(x)e^{i\mu x}dx \qquad (4.44)$$

where $\mu = kx_s/L$ is the spatial frequency. Using the spatial frequency, equation (4.44) becomes

$$U(\mu) = \int g(x)e^{i\mu x}dx. \qquad (4.45)$$

Therefore, the functions $U(\mu)$ and $g(x)$ constitute a one-dimensional Fourier transform pair. Notice that the spatial frequency has the dimension of wave number. Let us examine several cases of different aperture functions $g(x)$.

**Single-slit diffraction**: for a single slit of width $2b$, the aperture function is a single step function: $g(x) = 1$ for $-(b/2) < x < +(b/2)$; and $g(x) = 0$ otherwise as shown in figure 4.27. The corresponding diffraction integral is given by

$$U(\mu) = \int_{-\frac{b}{2}}^{+\frac{b}{2}} e^{i\mu x}dx = 2b\frac{\sin\left(\frac{1}{2}\mu b\right)}{\mu b}. \tag{4.46}$$

The normalized intensity $I/I_0$ of diffracted light is proportional to $|U(\mu)|^2$ (figure 4.28).

Since $\mu = kx_s/L = k\cdot\sin\theta$, we obtain

$$U(\theta) = 2b\frac{\sin\left(\frac{1}{2}kb\sin\vartheta\right)}{kb\sin\theta} = b\frac{\sin\beta}{\beta}$$

where $\beta = (1/2)kb\cdot\sin\theta$.

Therefore, the normalized diffraction intensity is given by

$$\frac{I}{I_0} = \left(\frac{\sin\beta}{\beta}\right)^2. \tag{4.47}$$

**Double-slit diffraction**: for a double-slit aperture shown in figure 4.29, each slit of width $b$ is separated by a distance $d$. The aperture function is thus given by $g(x) = 1$ for $0 < x < b$ and $h < x < d + b$; and $g(x) = 0$ otherwise.

The diffraction intensity is given by

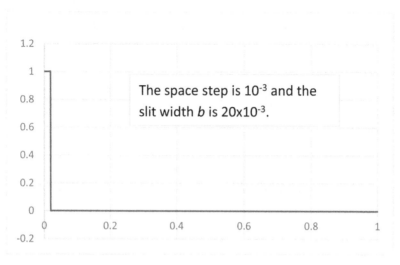

The space step is $10^{-3}$ and the slit width $b$ is $20\times10^{-3}$.

**Figure 4.27.** Single slit aperture function $g(x)$.

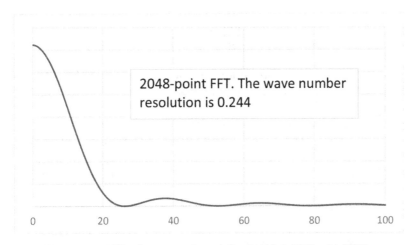

**Figure 4.28.** Diffraction pattern from a slit of width $b$ (2048-point FFT).

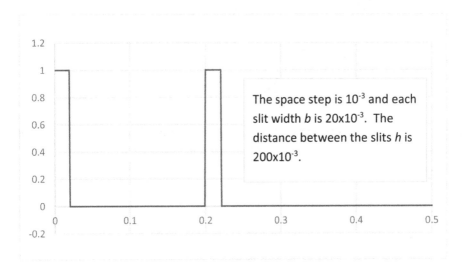

**Figure 4.29.** Double-slit aperture function.

$$U(\mu) = \int_0^{+\infty} g(x)e^{i\mu x}dx = \int_0^b e^{i\mu x}dx + \int_h^{h+b} e^{i\mu x}dx = \left(\frac{e^{i\mu b} - 1}{i\mu}\right)(1 + e^{i\mu h}). \quad (4.48)$$

Since $\mu = kx_s/L = k\cdot\sin\theta$ with the small angle approximation, the above equation becomes

$$U(\theta) = \left(\frac{e^{ikb\sin\theta} - 1}{ik\sin\theta}\right)(1 + e^{ikh\sin\theta}) = 2be^{i\beta}e^{i\gamma}\frac{\sin\beta}{\beta}\cos\gamma \quad (4.49)$$

where $\beta = (1/2)kb\cdot\sin\theta$ and $\gamma = (1/2)kh\cdot\sin\theta$. The intensity $I$ on the screen is proportional to $|U(\theta)|^2$:

$$I = \left(\frac{\sin \beta}{\beta}\right)^2 \cos^2 \gamma. \tag{4.50}$$

The factor $(\sin \beta/\beta)^2$ is the distribution function for a single slit. The term $\cos^2\gamma$ expresses the envelope for the interference fringes. Figure 4.30 depicts the diffraction intensity from the double slit.

**Multi-slit (diffraction grating) diffraction**: for a one-dimensional diffraction grating of $N$ identical parallel slits of width $b$ and separation (grating constant) $d$, the aperture function $g(x) = 1$ for $0 < x < b$, $h < x < d + b$, $2h < x < 2d + b$, ..., $(N-1)d < x < (N-1)d + b$; and $g(x) = 0$ otherwise. The diffraction integral is given by

$$
\begin{aligned}
U(\mu) &= \int_0^{+\infty} g(x)e^{i\mu x}dx \\
&= \int_0^b e^{i\mu x}dx + \int_h^{h+b} e^{i\mu x}dx + \int_{2h}^{2h+b} e^{i\mu x}dx + \cdots + \int_{(N-1)h}^{(N-1)h+b} e^{i\mu x}dx \\
&= \frac{e^{i\mu b} - 1}{i\mu}[1 + e^{i\mu h} + e^{2i\mu h} + \cdots + e^{i(N-1)\mu b}] \\
&= \frac{e^{i\mu b} - 1}{i\mu} \cdot \frac{1 - e^{i\mu Nh}}{1 - e^{i\mu h}} = \frac{e^{ikb \sin \theta}}{ik \sin \theta} \cdot \frac{1 - e^{ikNh \sin \theta}}{1 - e^{ikh \sin \theta}} \\
&= be^{i\beta}e^{i(N-1)\gamma}\left(\frac{\sin \beta}{\beta}\right)\left(\frac{\sin N\gamma}{\sin \gamma}\right)
\end{aligned}
\tag{4.51}
$$

where $\mu = k \sin \theta$, $\beta = (1/2)kb \cdot \sin\theta$ and $\gamma = (1/2)kh \cdot \sin\theta$. Since the intensity $I$ on the screen is proportional to $|U(\theta)|^2$, we obtain

The envelope is the diffraction pattern due to a single slit of the same slit width $b$, i.e., $(\sin\beta/\beta)^2$-factor. The fine pattern is $\cos^2\gamma$.

**Figure 4.30.** Diffraction pattern of a double slit (2048-point FFT).

$$I = I_0 \left( \frac{\sin \beta}{\beta} \right)^2 \left( \frac{\sin N\gamma}{N \sin \gamma} \right)^2. \tag{4.52}$$

Figures 4.31 and 4.32 show the multiple-slit aperture function and its diffraction pattern, respectively.

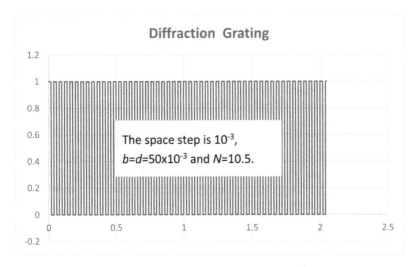

**Figure 4.31.** Aperture function $g(x)$ of 10-slit aperture.

**Figure 4.32.** Diffraction pattern of multiple-slit ($N = 102$ and 2048-point FFT).

### 4.4.2 Two-dimensional diffraction

Let us start with the Huygens' principle—the *secondary spherical waves* at each point of an aperture, and superposition of the secondary waves form the diffraction pattern. Consider the coordinates of an apertured plane and a screen as shown in figure 4.33 where $g(x, y)$ is a two-dimensional aperture function.

The amplitude of a diffraction pattern observed at point $P(x_s, y_s)$ on the screen is given by two-dimensional integral over the aperture:

$$U_P(x_s, y_s) = \int_{\text{apature}} \frac{C \exp(ikr)}{i\lambda r} g(x, y)dxdy \tag{4.53}$$

where $C$ is a constant, $r$ is the distance from a point inside the aperture to the observing point on the screen, $g(x, y)$ is the two-dimensional aperture function where $g(x,y) = 1$ if $(x,y)$ is inside the aperture; and 0 otherwise. The integral element of $dxdy$ is the area element of the aperture. The integrant is essentially the secondary spherical waveform in the aperture.

When the distance from the light source to the aperture plane is very long, and the distance from the aperture to the observation screen—denoted as $R$—is also much larger than the aperture size, the incident and diffracted spherical waves have negligible curvatures. In this case, the angular spread of the diffracted light is small, and we can regard both the incident and diffracted waves as *plane waves*. This is

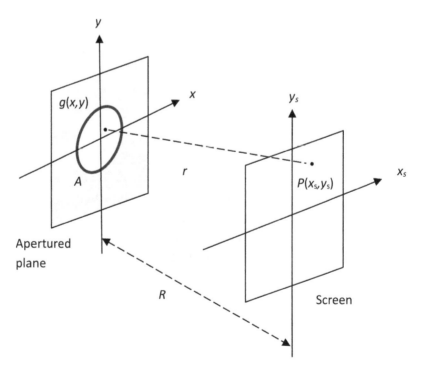

**Figure 4.33.** Configuration of an apertured plane and a screen for observing diffraction.

called the Fraunhofer diffraction, and the Huygens' principle can be expressed by the following equation:

$$U_P(x_s, y_s) = \frac{C}{i\lambda R} \int_{apature} g(x, y)\exp(ikr)dxdy. \tag{4.54}$$

Now, we apply the 'Fresnel approximation' to equation (4.54). Because $R$ is considerably large it is denotated as $(x-x_s)$ and $(y-y_s)$, the distance $r$ can be approximated to be

$$r = \sqrt{R^2 + (x - x_s)^2 + (y - y_s)^2} = R\sqrt{1 + \frac{(x - x_s)^2 + (y - y_s)^2}{R^2}} \tag{4.55}$$

$$\approx R + \frac{1}{2R}[(x - x_s)^2 + (y - y_s)^2] = R + \frac{1}{2R}[(x^2 + y^2) + (x_s^2 + y_s^2)] - \frac{1}{R}(xx_s + yy_s).$$

Furthermore, because the aperture size is very small, we may drop the $(x^2+y^2)$ term from equation (4.55) to get

$$r = R + \frac{1}{2R}[(x_s^2 + y_s^2)] - \frac{1}{R}(xx_s + yy_s), \tag{4.56}$$

whence the diffraction pattern $U_P$ becomes

$$U_P(x_s, y_s) = \frac{C}{i\lambda R}e^{ikR}e^{i\frac{k}{2R}(x_s^2+y_s^2)} \int_{-\infty}^{\infty}\int_{-\infty}^{\infty} g(x, y)\exp\left[-\frac{ik}{R}(xx_s + yy_s)\right]dxdy$$

$$= C' \int_{-\infty}^{\infty}\int_{-\infty}^{\infty} g(x, y)\exp\left[-\frac{ik}{R}(x_sx + y_sy)\right]dxdy \tag{4.57}$$

where $C' = (C/i\lambda R)e^{ikR}e^{i\frac{k}{2R}(x_s^2+y_s^2)}$. Equation (4.57) indicates that the diffraction pattern, $U_p$, is the two-dimensional Fourier transform of the geometrical shape of the aperture function $g(x, y)$. We describe two examples of two-dimensional diffraction patterns below.

**Rectangular aperture**: if the aperture is a rectangle of size $D_x$ by $D_y$,

$$U_P(x_s, y_s) = C' \int_{-D_x/2}^{D_x/2}\int_{-D_y/2}^{D_y/2} g(x, y)\exp\left[-\frac{ik}{R}(xx_s + yy_s)\right]dxdy$$

$$= C'D_xD_y\frac{\sin\left(\frac{kD_x}{2R}x_s\right)\sin\left(\frac{kD_y}{2R}y_s\right)}{\left(\frac{kD_x}{2R}\right)x_s\left(\frac{kD_y}{2R}\right)y_s}, \tag{4.58}$$

where the area integral is very much the same as equation (4.46) of the one-dimensional slit. Figure 4.34 shows the intensity of the diffraction pattern through a square aperture, calculated by equation (4.62), where $C' = 1$, $D_x = D_y = 1$, and $k/2R = 1$. For this drawing, we used EXCEL's [Surface Chart] option to draw figure 4.34. Refer to appendix A13.3 for the drawing procedure. Figure 4.35 simulates the intensity of diffraction from an arbitrary square aperture (a 9×9 pixels

Rotated view           Top view

**Figure 4.34.** Intensity of diffraction pattern through a square aperture.

2D-FFT of square aperture

**Figure 4.35.** Intensity of FFT pattern of a square step.

aperture on a plane of 512×512 pixels of height 1) using the two-dimensional FFT code of appendix A13.5. Only 255 × 255 points that EXCEL can display are shown from the acquired FFT image.

**Circular aperture**: if the aperture is a circle of diameter $D$, we use the polar coordinates defined as $\{x = \rho\cos\theta,\ y = \rho\sin\theta\}$ and $\{x_s = \rho_s\cos\theta_s,\ y_s = \rho_s\sin\theta_s\}$.

$$U_P(x_s,\ y_s) = C' \int_0^{D/2} \rho\,d\rho \int_0^{2\pi} d\theta \exp\left[-\frac{ik,\ y}{R}\rho\rho_s \cos(\theta - \theta_s)\right]$$

$$= 2\pi C' \int_0^{D/2} \rho\,d\rho J_0\!\left(\frac{k}{R}\rho\rho_s\right) = \pi C'\!\left(\frac{D}{2}\right)^2 \frac{2J_1\!\left(\frac{kD}{2R}\rho_s\right)}{\left(\frac{kD}{2R}\rho_s\right)}$$

(4.59)

where the Bessel function $J_n(x)$

$$J_n(x) = \frac{1}{2\pi(i)^n} \int_0^{2\pi} e^{ix\cos\theta} e^{in\theta} d\theta \tag{4.60}$$

and the recursion relation

$$\frac{d}{dx}[x^{n+1}J_{n+1}(x)] = x^n J_n(x) \tag{4.61}$$

are used for the integral calculation. Figure 4.36 shows the intensity of the diffraction pattern through a circular aperture, computed by equation (4.63) where $C' = 1/\pi$, $(D/2)^2 = 0.5$, and $kD/R = 1$. We used the built-in Bessel function in EXCEL. Refer to appendix A13.3 for the drawing procedure.

Rotated view          Top view

**Figure 4.36.** Diffraction intensity pattern through a circular aperture.

### 4.4.3 Lens

Imagine a *point* light source is placed in front of a lens as shown in figure 4.37. The light wave coming to the lens is an incoming spherical wave:

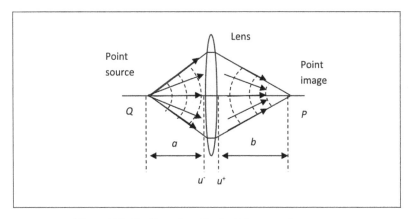

**Figure 4.37.** Configuration of point light source-lens-screen.

$$U_{\text{before}}(x_\ell, y_\ell) = C \exp\left[i\frac{\pi}{\lambda a}(x_\ell^2 + y_\ell^2)\right]. \tag{4.62}$$

The light wave immediately after the lens to form the image point at $P$ can be considered as the 'backward' spherical wave from point $P$ to the lens:

$$U_{\text{after}}(x_\ell, y_\ell) = C' \exp\left[i\frac{\pi}{\lambda b}(x_\ell^2 + y_\ell^2)\right] \cdot p(x_\ell, y_\ell). \tag{4.63}$$

In the above equations, $a$ and $b$ are the distances from the light source to the lens and from the lens to the image at point $P$, satisfying the lens formula, $(1/a) + (1/b) = (1/f)$. The function $p(x_\ell, y_\ell)$ in the above equations is called a pupil function: $p(x_\ell, y_\ell) = 1$ inside the lens; and $= 0$, otherwise.

We define the transmissivity of the lens denoted as $t(x_\ell, y_\ell) = U_{\text{after}}/U_{\text{before}}$. Here, we assume no change in the amplitude before and after the lens, i.e., $C = C'$ and thus the transmissivity actually expresses the phase change of the light while passing through the lens [8]. Using the lens formula, we obtain

$$t(x_\ell, y_\ell) = \frac{p(x_\ell, y_\ell)\exp[-i\frac{\pi}{\lambda b}(x_\ell^2 + y_\ell^2)]}{\exp[i\frac{\pi}{\lambda a}(x_\ell^2 + y_\ell^2)]} = p(x_\ell, y_\ell)\exp[-i\frac{\pi}{\lambda}\left(\frac{1}{a} + \frac{1}{b}\right)(x_\ell^2 + y_\ell^2)]$$

$$= p(x_\ell, y_\ell)\exp[-i\frac{k}{2f}(x_\ell^2 + y_\ell^2)]. \tag{4.64}$$

Now, as shown in figure 4.38, we place an object characterized by the object function, $g(x_o, y_o)$, at a distance $a$ in front of the lens where $(x_o, y_o)$ are the object coordinates, and observe its image on the screen placed at distance $b$ from the lens.

The light wave from the object that hits the lens is given by equations (4.62). Using the Fresnel approximation applied in equations (4.55)–(4.57), we obtain

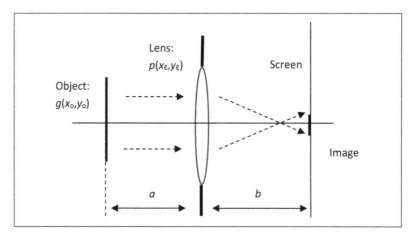

**Figure 4.38.** Schematic diagram of object-lens-screen configuration.

$$U_{\text{before}}(x_\ell, y_\ell) = \frac{C}{i\lambda a} \iint_{\text{object}} g(x_o, y_o)\exp(ikr)dx_ody_o$$

$$= \frac{C}{i\lambda a} \iint_{\text{object}} g(x_o, y_o)\exp\left[ik\left\{a + \frac{1}{2a}((x_\ell - x_o)^2 + (y_\ell - y_o)^2)\right\}\right]dx_ody_o$$

$$= \frac{C}{i\lambda a}e^{ika} \iint_{\text{object}} g(x_o, y_o)\exp\left[-i\frac{k}{2a}((x_\ell - x_o)^2 + (y_\ell - y_o)^2)\right]dx_ody_o$$

$$= \frac{C}{i\lambda a}e^{ika}e^{i\frac{k}{2a}(x_\ell^2 + y_\ell^2)} \iint_{\text{object}} g(x_o, y_o)\exp\left[-i\frac{k}{2a}(x_\ell x_o + y_\ell y_o)\right]dx_ody_o.$$

(4.65)

The light wave immediately after the lens $(x_\ell, y_\ell)$ is given by $U_{\text{after}}(x_\ell, y_\ell) = t(x_\ell, y_\ell)\cdot U_{\text{before}}(x_\ell, y_\ell)$. Thus, the formed image, $U_{\text{image}}(x_s, y_s)$, on screen $(x_s, y_s)$ should be given by

$$\text{image}(x_s, y_s) = \frac{C}{i\lambda b} \iint_{\text{lens}} U_{\text{after}}(x_\ell, y_\ell)\exp\left[ik\left\{b + \frac{1}{2b}(x_s - x_\ell)^2 + (y_s - y_\ell)^2\right\}\right]dx_\ell dy_\ell$$

$$= \frac{C}{i\lambda b}e^{ikb}e^{i\frac{k}{2b}(x_s^2 + y_s^2)} \iint_{\text{lens}} U_{\text{before}}(x_\ell, y_\ell)t(x_\ell, y_\ell)\exp\left[-i\frac{k}{2b}(x_\ell x_s + y_\ell y_s)\right]dx_\ell dy_\ell$$

$$= \frac{C}{i\lambda b}e^{ikb}e^{i\frac{k}{2b}(x_s^2 + y_s^2)} \iint_{\text{lens}} \left[\frac{C}{i\lambda a}e^{ika}e^{i\frac{k}{2a}(x_\ell^2 + y_\ell^2)} \iint_{\text{object}} g(x_o, y_o)\exp\left[-i\frac{k}{2a}(x_\ell x_o + y_\ell y_o)\right]dx_ody_o\right]$$

$$\times p(x_\ell, y_\ell)\exp\left[-i\frac{k}{2f}(x_\ell^2 + y_\ell^2)\right]\exp\left[-i\frac{k}{2b}(x_\ell x_s + y_\ell y_s)\right]dx_\ell dy_\ell$$

(4.66)

$$= -\frac{C^2}{\lambda^2 ab}e^{ik(a+b)}e^{i\frac{k}{2b}(x_s^2 + y_s^2)} \iint_{\text{lens}} \left[\iint_{\text{object}} g(x_o, y_o)\exp\left\{-i\frac{k}{2a}(x_\ell x_o + y_\ell y_o)\right\}dx_ody_o\right]$$

$$\times p(x_\ell, y_\ell)e^{i\frac{k}{2a}(x_\ell^2 + y_\ell^2) - i\frac{k}{2f}(x_\ell^2 + y_\ell^2)}\exp\left\{-i\frac{k}{2b}(x_\ell x_s + y_\ell y_s)\right\}dx_\ell dy_\ell$$

$$= C_{\text{sys}} \iint_{\text{lens}} \left[\iint_{\text{object}} g(x_o, y_o)\exp\left\{-i\frac{k}{2a}(x_\ell x_o + y_\ell y_o)\right\}dx_ody_o\right]$$

$$\times q(x_\ell, y_\ell)\exp\left\{-i\frac{k}{2b}(x_\ell x_s + y_\ell y_s)\right\}dx_\ell dy_\ell$$

where $C_{\text{sys}}$ is the coefficient in front of the integral and $q(x_\ell, y_\ell) = p(x_\ell, y_\ell)\exp[-ik(x_\ell^2, y_\ell^2)/2b]$. The integral is the convoluted Fourier transform of the object function $g(x_o, y_o)$, and the pupil function $q(x_\ell, y_\ell)$. For detailed discussion of lens optics, refer to the reference [8].

### 4.4.4 Hologram

If we can record an optical image, $F(x,y) = f(x,y)e^{i\varphi(x,y)}$, in the wave number domain, we can reproduce, $F(x,y)$, in the spatial domain by performing a Fourier transform. However, conventional photographs can only record the intensity, $|F(x,y)|^2$, losing the phase information and the image is 'flat' without depth information.

Holography can make a photograph that restores the phase information in the following steps. First, take a photograph of $F(x,y)$ with a superposed reference light $A(x,y)$. The photographed image will be a record of the intensity of the superposed light

$$|A(x, y) + F(x, y)|^2 = |A(x, y)|^2 + |F(x, y)|^2$$
$$+ A^*(x, y)F(x, y) + A(x, y)F^*(x, y). \tag{4.67}$$

Next, place the photograph at the object position and irradiate it with the reference light $A(x,y)$. The amplitude of the transmitted light is now given by

$$A(x, y)|A(x, y) + F(x, y)|^2 = A(x, y)|A(x, y)|^2$$
$$+ A(x, y)|F(x, y)|^2 + |A(x, y)|^2 F(x, y) + A^2(x, y)F^*(x, y). \tag{4.68}$$

If we can pick out the third term, $|A(x,y)|^2 F(x,y)$, we will be able to reconstruct $F(x, y)$! This is the principle of how we get a hologram.

In practice, for the transmission hologram, the incident direction of the superposed reference light is tilted by angle $\theta$ to the $x$-axis, i.e. $A(x,y) = a_0 e^{ikx\sin\theta}$ where $k = 2\pi/\lambda$. The intensity of the superposed photograph image is given by

$$I(x, y) = |F(x, y) + A(x, y)|^2 = |f(x, y)e^{i\varphi(x, y)} + a_0 e^{ik \, x \, \sin\theta}|^2$$
$$= |f(x, y)|^2 \quad + |a_0|^2 + 2a_0 f(x, y)\cos[\varphi(x, y) - kx \sin\theta]. \tag{4.69}$$

By recording this intensity on a 'transparent' film, the distribution of the transmitted light will be proportional to $I(x,y)$ in addition to a 'background noise' level $t_0$: $t(x, y) = t_0 + \alpha I(x,y)$ where $\alpha$ is the proportional constant. Irradiate the film with the same reference light, $A(x,y)$, and we obtain

$$A(x, y)t(x, y) = [t_0 + \alpha |f(x, y)|^2 + \alpha |a_0|^2]a_0 e^{ik \, x \, \sin\theta}$$
$$+ \alpha a_0^2 f(x, y)e^{i\varphi(x, y)} + \alpha f(x, y)a_0^2 e^{-i\varphi(x, y)}e^{i2k \, x \, \sin\theta}. \tag{4.70}$$

The second term on the right-hand-side of the above equation can reconstruct $F(x, y) = f(x, y)e^{i\varphi(x,y)}$. Figure 4.39 depicts a schematic diagram of hologram reconstruction.

**Figure 4.39.** Hologram reconstruction.

### 4.4.5 X-ray diffraction

As we have seen, a diffraction pattern is the Fourier transform of aperture shape. This is also true for diffractions (scatterings) of x-ray, electrons, and neutrons

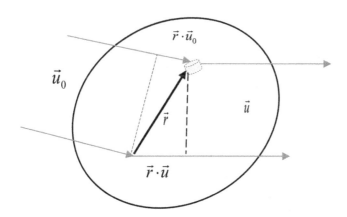

**Figure 4.40.** X-ray scattering by electrons in an object.

through target structures. Therefore, by performing the Fourier inverse transform of the diffraction patterns, the object structure can be computed.

First, we consider x-ray scattering (diffraction) by a volume of elements of an object interacting with x-rays. Figure 4.40 is a schematic diagram of the x-ray scattering. The incident direction of a monochrome x-ray of wavelength $\lambda_0 = 2\pi/k_0$ where $k_0$ is the wave number is represented by unit vector $\vec{u}_0$. The scattered x-ray has the wavelength and the wave number given by $\lambda = 2\pi/k$ and in the direction represented by $\vec{u}$. The path difference between the incident and the scattered x-ray is given by $\delta = \vec{r} \cdot (\vec{u} - \vec{u}_0)$ or the phase difference is expressed as $(2\pi/\lambda)(\vec{u} - \vec{u}_0) = 2\pi(\vec{k} - \vec{k}_0) \cdot \vec{r}$. Using the electron density $\rho(\vec{r})$ at the volume element of the scatterer $dV$ the amplitude of the scatted x-ray from the volume element $dV$ should be given by $I_e \cdot [\rho(\vec{r})\exp(2\pi i(\vec{k} - \vec{k}_0) \cdot \vec{r})dV]$ where $I_e$ is the intensity of the scattered x-ray from a single electron. The intensity of the scattered x-ray from the entire scatterer can be given by

$$I = I_e \cdot |\int_{\text{object}} \rho(\vec{r})\exp[2\pi i(\vec{k} - \vec{k}_0) \cdot \vec{r}]dV|^2. \tag{4.71}$$

This is essentially the Fourier transform of the electron distribution function $\rho(\vec{r})$.

Next, when the target object is a single crystal, the position vector $\vec{r}$ can be expressed with lattice unit vectors $(\vec{a}_1, \vec{a}_2, \vec{a}_3)$: $\vec{r} = m_1\vec{a}_1 + m_2\vec{a}_2 + m_3\vec{a}_3$ where $\{m_i; i = 1, 2, 3\}$ are integers to indicate a lattice point $(m_1, m_2, m_3)$ of the crystal. Thus,

$$\begin{aligned}I_{\text{crystal}} &= I_e \cdot |\int_{\text{crytal}} \rho(\vec{r})\exp[2\pi i(\vec{k} - \vec{k}_0) \cdot \vec{r}]dV|^2 \\ &= I_e |\sum_{m_1}\sum_{m_2}\sum_{m_3} \exp(2\pi i(\vec{k} - \vec{k}_0) \cdot (m_1\vec{a}_1 + m_2\vec{a}_2 + m_3\vec{a}_3)) \int_{\text{unitcell}} \rho(\vec{r}')\exp[2\pi i(\vec{k} - \vec{k}_0) \cdot \vec{r}']dV'|^2\end{aligned} \tag{4.72}$$

where $\vec{r}'$ is the position vector from the lattice point $(m_1, m_2, m_3)$ to a point inside the unit cell, and $\vec{r} = m_1\vec{a}_1 + m_2\vec{a}_2 + m_3\vec{a}_3 + \vec{r}'$ is the position vector from the origin. Because of the periodic property of a single crystal, the electron distribution function is the same in any unit cell: $\rho(\vec{r}) = \rho(\vec{r}')$.

We now define the Laue function as

$$G(\vec{k} - \vec{k}') \equiv \sum_{m_1} \sum_{m_2} \sum_{m_3} \exp(2\pi i (\vec{k} - \vec{k}_0) \cdot (m_1 \vec{a}_1 + m_2 \vec{a}_2 + m \vec{a}_3))$$

$$= \sum_{m_1=0}^{L-1} \exp(2\pi i m_1 (\vec{k} - \vec{k}_0) \cdot \vec{a}_1) \sum_{m_2=0}^{M-1} \exp(2\pi i m_2 (\vec{k} - \vec{k}_0) \cdot \vec{a}_3) \sum_{m_3=0}^{L-1} \exp(2\pi i m_3 (\vec{k} - \vec{k}_0) \cdot \vec{a}_3) \quad (4.73)$$

$$= \frac{\sin[\pi L((\vec{k} - \vec{k}_0) \cdot \vec{a}_1)]}{\sin[\pi((\vec{k} - \vec{k}_0) \cdot \vec{a}_1)]} \frac{\sin[\pi M((\vec{k} - \vec{k}_0) \cdot \vec{a}_2)]}{\sin[\pi((\vec{k} - \vec{k}_0) \cdot \vec{a}_2)]} \frac{\sin[\pi N((\vec{k} - \vec{k}_0) \cdot \vec{a}_3)]}{\sin[\pi((\vec{k} - \vec{k}_0) \cdot \vec{a}_3)]}$$

where we assume the single crystal has the size of $(La_1, Ma_2, Na_3)$. The intensity of the diffracted x-ray from the crystal is given by

$$I_{\text{crystal}} = I_e \, |G(\vec{k} - \vec{k}')|^2 \cdot \left| \int_{\text{unitcell}} \rho(\vec{r}') \exp[2\pi i (\vec{k} - \vec{k}_0) \cdot \vec{r}'] dV'^2 \right|. \quad (4.74)$$

If $L$, $M$, and $N$ are sufficiently large, the Laue function is known to have sharp peaks at $(\vec{k} - \vec{k}_0) \cdot \vec{a}_1 = p_1$, $(\vec{k} - \vec{k}_0) \cdot \vec{a}_2 = p_2$, and $(\vec{k} - \vec{k}_0) \cdot \vec{a}_3 = p_3$ where $\{p_i; i = 1, 2, 3\}$ are 0 or integers [9].

If we introduce the reciprocal lattice vectors:

$$\vec{b}_1 = 2\pi \frac{\vec{a}_2 \times \vec{a}_3}{\vec{a}_1 \cdot (\vec{a}_2 \times \vec{a}_3)}, \quad \vec{b}_2 = 2\pi \frac{\vec{a}_3 \times \vec{a}_1}{\vec{a}_1 \cdot (\vec{a}_2 \times \vec{a}_3)}, \quad \text{and} \quad \vec{b}_3 = 2\pi \frac{\vec{a}_1 \times \vec{a}_2}{\vec{a}_1 \cdot (\vec{a}_2 \times \vec{a}_3)} \quad (4.75)$$

where $\vec{a}_j \cdot \vec{b}_k = 2\pi \delta_{jk}$, the Laue condition satisfies $\vec{k} - \vec{k}' = p_1 \vec{b}_1 + p_2 \vec{b}_2 + p_3 \vec{b}_3$ [10].

## 4.5 Stochastic processes and Fourier transform

### 4.5.1 Characteristic function and central limit theorem

Physics phenomena are often stochastic where Fourier transforms play important roles. We start with the central limit theorem as we often apply it. The central limit theorem states that any probability distribution will approach Gaussian with sufficient numbers of independent trials [11]. Suppose $n$-independent values, $\{x_j; j = 1, 2, \ldots, n\}$ that follow a probability distribution function, $f(x)$, we observe to compute their mean and the standard deviation. For simplicity, we assume the mean is zero, then

$$X = \sum_{j=1}^{n} x_j$$

$$\langle X \rangle = 0$$

$$\sigma_j \equiv \sqrt{\langle x_j^2 \rangle - \langle x_j \rangle^2} = \sqrt{\langle x_j^2 \rangle} \quad (4.76)$$

$$s^2 \equiv \sum_{j=1}^{n} \sigma_j^2.$$

The mathematical expression of the central limit theorem is that a new variable, $Y = X/s$ follows the Gaussian distribution of the mean, $\langle Y \rangle = 0$, and the standard deviation, $\sigma_Y = 1$:

$$f(Y) = \frac{1}{\sqrt{2\pi}} \exp\left(-\frac{1}{2}Y^2\right). \tag{4.77}$$

To confirm the central limit theorem, we define the characteristic function, $\Phi$ ($\xi$) $= \langle \exp(i\xi x) \rangle$, which is the Fourier transform of the probability distribution function $f(x)$:

$$\Phi(\xi) = \int_{-\infty}^{\infty} e^{i\xi x} f(x)dx. \tag{4.78}$$

From the $m$th moments, $\langle x^m \rangle = \int x^m f(x)dx$, the characteristic function can be written as

$$\Phi(\xi) = \sum_{m=0}^{\infty} \frac{\langle i\xi \rangle^m}{m!} \langle x^m \rangle. \tag{4.79}$$

Conversely, the $m$th moments can be generated from the characteristic function:

$$\langle x^m \rangle = \left(\frac{1}{i}\right)^m \left[\left(\frac{d}{d\xi}\right)^m \Phi(\xi)\right]_{\xi=0}. \tag{4.80}$$

For a Gaussian distribution function

$$f(x) = \frac{1}{\sqrt{2\pi\sigma^2}} \exp\left(-\frac{(x - \langle x \rangle)^2}{2\sigma^2}\right), \tag{4.81}$$

its characteristic function becomes

$$\Phi(\xi) = \int_{-\infty}^{\infty} e^{i\xi x} f(x)dx = \frac{1}{\sqrt{2\pi\sigma^2}} \int_{-\infty}^{\infty} \exp\left[-\frac{(x - \langle x \rangle)^2}{2\sigma^2} + ix\xi\right]$$

$$= \frac{1}{\sqrt{2\pi\sigma^2}} \int_{-\infty}^{\infty} \exp\left[i\langle x \rangle\xi - \frac{\sigma^2}{2}\xi - \frac{(x - \langle x \rangle - i\sigma^2\xi)^2}{2\sigma^2}\right]dx \tag{4.82}$$

$$= \exp\left[i\langle x \rangle\xi - \frac{\sigma^2}{2}\xi\right].$$

Now, we define cumulant function, $\Psi(\xi)$, as $\Phi(\xi) = \exp[\Psi(\xi)]$, or

$$\Psi(\xi) = \ln \Phi(\xi) = \sum_{m=1}^{\infty} \frac{(i\xi)^m}{m!} \langle x^m \rangle_c, \tag{4.83}$$

where $\langle x^m \rangle_c$ is called the $m$th cumulant. Notice that $\langle x \rangle_c = \langle x \rangle$ and $\langle x^2 \rangle_c = \langle x^2 \rangle - \langle x \rangle^2$, and the Gaussian distribution has no $m$th cumulant of $m \geqslant 3$, i.e. $\langle x^m \rangle_c = 0$ if $m \geqslant 3$. This is a distinct property of the Gaussian distribution.

Suppose the cumulant functions of $x_j$ ($j= 1,2, ..., n$) are $\psi_j(\xi)$ and those of $X$ are $\Psi$ $(\xi)$,

$$\Psi(\xi) = \sum_{j=1}^{n} \psi_j(\xi). \tag{4.84}$$

Thus, if $\psi_j(\xi) = \sum_{m=1}^{\infty} \frac{(i\xi)^m}{m!} \langle x^m \rangle_c$, then we obtain

$$\Psi(\xi) = \sum_{j=1}^{n} \psi_j(\xi) = i\xi \sum_{j=1}^{n} \langle x_j \rangle - \frac{\xi^2}{2} \sum_{j=1}^{n} \sigma_j^2 + \frac{(i\xi)^3}{3!} \sum_{j=1}^{n} \langle x_j^3 \rangle_c + \cdots$$

$$= -\frac{\xi^2}{2} s_n^2 + \frac{(i\xi)^3}{3!} \sum_{j=1}^{n} \langle x_j^3 \rangle_c + \cdots, \tag{4.85}$$

and the characteristic function of $Y$ is given by

$$\Psi_Y(\eta) = \langle e^{i\eta Y} \rangle = \langle e^{i\eta X/s} \rangle = \exp\left[ -\frac{1}{2}\eta^2 + \frac{(i\eta)^3}{3!} \sum_{j=1}^{n} \frac{\langle x_j^3 \rangle_c}{s^3} + \cdots \right]. \tag{4.86}$$

Assuming that the $m$th moments of $\{x_j\}$ are all finite and similar, only the first term of the above equation remains, whence we obtain

$$\Psi_Y(\eta) = \langle e^{i\eta Y} \rangle \rightarrow \exp\left[ -\frac{1}{2}\eta^2 \right]. \tag{4.87}$$

Therefore, as the central limit theorem states, the probability distribution function of $Y$ approaches the Gaussian.

### 4.5.2 Brownian motion

One-dimensional Brownian motion can be described by the Langevin equation given by

$$m\frac{dv(t)}{dt} = -m\gamma v(t) + f(t) \tag{4.88}$$

where $m$, $\gamma$, and $f(t)$ represent the mass of a Brownian particle (e.g. pollen), viscosity (or friction), and random force due to collisions with surrounding particles such as water molecules, respectively [12]. From equation (4.88), we obtain the ensemble average of the correlation function of velocities defined as

$$\left\langle \frac{dv(t)}{dt} v(0) \right\rangle = -\gamma\langle v(t)v(0) \rangle + \frac{1}{m}\langle f(t)v(0) \rangle \text{ for } t > 0. \tag{4.89}$$

Because the Brownian motion is a Markov process, there is no correlation between two successive collisions, $\langle v(t_1)f(t_2) \rangle = 0$ where $t_1$ and $t_2$ are different times. Thus,

$$\left\langle \frac{dv(t)}{dt}v(0) \right\rangle = \frac{d}{dt}\langle v(t)v(0)\rangle = -\gamma\langle v(t)v(0)\rangle. \tag{4.90}$$

This equation can be solved to obtain the correlation function

$$\langle v(t)v(0)\rangle = \langle v(0)^2\rangle e^{-\gamma t}. \tag{4.91}$$

The Brownian particles should be in thermal equilibrium with the surrounding medium at temperature $T$, and the classical equipartition law, $\langle (1/2)mv^2\rangle = (1/2)k_BT$, is applicable where $k_B$ is the Boltzmann constant. Therefore

$$\langle v(t)v(0)\rangle = \frac{k_BT}{m}e^{-\gamma t}. \tag{4.92}$$

The Einstein reaction can be derived using the above equation. Assume that the initial positions of the Brownian particles are at the origin ($x=0$ at $t=0$), the position of the Brownian particle is given by $x(t) = \int_0^t v(\tau)d\tau$, and

$$\langle x(t)^2\rangle = \int_0^t\int_0^t v(\tau)v(\tau')d\tau d\tau' = \frac{k_BT}{m}\int_0^t\int_0^t e^{-\gamma(\tau-\tau')}d\tau d\tau'$$

$$= \frac{2k_BT}{m}\left(\frac{t}{\gamma} - \frac{1-e^{-\gamma t}}{\gamma^2}\right) = \frac{2k_BT}{m}\left(\frac{t}{\gamma} - \frac{1-(1-\gamma t+(1/2)(\gamma t)^2+\cdots)}{\gamma^2}\right) \tag{4.93}$$

after a sufficiently long time, $t\gg 1/\gamma$ and terms of $t^2$ or higher can be neglected, whence we obtain the Einstein relation, which states that fluctuation is proportional to time [13]:

$$\langle x(t)^2\rangle = \left(\frac{4k_BT}{\gamma m}\right)t. \tag{4.94}$$

### 4.5.3 Fluctuation–dissipation theorem in Brownian motion

We show that equation (4.93) has the very special meaning that a linear response of a system in thermodynamic equilibrium to a 'small' external perturbation can be expressed in terms of spontaneous fluctuation in the thermodynamic equilibrium state. Let us start with the Langevin equation (4.89) to explain this statement.

Take the Fourier transforms of Langevin equation (4.89), and we obtain

$$i\omega m V(\omega) = -m\gamma V(\omega) + F(\omega) \text{ and thus } V(\omega) = \frac{1}{m}\frac{F(\omega)}{i\omega + \gamma} = G(\omega)F(\omega) \tag{4.95}$$

where $V(\omega)$ and $F(\omega)$ are Fourier transforms of $v(t)$ and $f(t)$:

$$V(\omega) = \int_{-\infty}^{\infty} v(t)e^{i\omega t}dt \text{ and } F(\omega) = \int_{-\infty}^{\infty} f(t)e^{i\omega t}dt, \tag{4.96}$$

and $G(\omega) = 1/(i\omega+\gamma)$ is the transfer function defined in the linear response theory. Because the transfer function $G(\omega)$ is the Fourier transform of the impulse response $g$ ($t$) (section 2.3.3), $g(t)$ can be given by the inverse Fourier transform of $G(\omega)$.

Let us obtain the impulse response $g(t)$, which should satisfy the causality.

$$G(\omega) = \frac{1}{m} \frac{1}{\gamma + i\omega} = \int_{-\infty}^{\infty} g(t)e^{-i\omega t}dt. \tag{4.97}$$

Using the integral $\int_{0}^{\infty} e^{-\gamma t}e^{-i\omega t}dt = \frac{1}{\gamma + i\omega}$, we obtain

$$G(\omega) = \frac{1}{m} \frac{1}{\gamma + i\omega} = \frac{1}{m}\int_{0}^{\infty} e^{-\gamma t}e^{-i\omega t}dt = \int_{0}^{\infty} g(t)e^{-i\omega t}dt \tag{4.98}$$

Thus, from equation (4.98), the impulse response is given by

$$g(t) = \frac{1}{m}e^{-\gamma t} = \frac{\langle v(t)v(0)\rangle}{\langle v(0)^2\rangle}. \tag{4.99}$$

According to equation (4.99), the impulse response indicates the energy dissipation, which is expressed by $e^{-\gamma t}$, and the response function can be expressed by the correlation function—$\langle v(t)v(0)\rangle$—in the thermal equilibrium state. This is a fluctuation–dissipation theorem [14].

The fluctuation–dissipation relation can be also derived from the Winer–Khintchine theorem (appendix A7). The power spectrum of $V(\omega)$—denoted as $S_V(\omega)$—is given by the power spectrum of $F(\omega) - S_F(\omega)$.

$$S_V(\omega) = \frac{1}{m^2} \frac{1}{|i\omega + \gamma|^2}S_F(\omega) = \frac{1}{m^2}\frac{S_F(\omega)}{\omega^2 + \gamma^2}. \tag{4.100}$$

According to the Winer–Khintchine theorem, the power spectrum, $S_V(\omega)$, can be expressed by the Fourier transform of correlation function:

$$S_V(\omega) = \int_{-\infty}^{\infty} \langle v(t_0)v(t_0 + \tau)\rangle e^{-i\omega \tau}d\tau. \tag{4.101}$$

Thus, we obtain

$$S_V(\omega) = \int_{-\infty}^{\infty} \langle v(t_0)v(t_0 + \tau)\rangle e^{-i\omega \tau}d\tau = \frac{1}{m^2}\frac{S_F(\omega)}{\omega^2 + \gamma^2}. \tag{4.102}$$

The inverse Fourier transform of the above equation gives

$$\langle v(t_0)v(t_0 + \tau)\rangle = \frac{1}{2\pi}\int_{-\infty}^{\infty} \frac{1}{m^2}\frac{S_F(\omega)}{\omega^2 + \gamma^2}e^{i\omega \tau}d\omega. \tag{4.103}$$

Assuming that the power spectrum of the random force is 'white', or $S_F(\omega) = \text{constant} = S_0$, equation (4.104) becomes

$$\begin{aligned}
\langle v(t_0)v(t_0 + \tau)\rangle &= \frac{1}{2\pi}\int_{-\infty}^{\infty} \frac{1}{m^2}\frac{S_0}{\omega^2 + \gamma^2}e^{i\omega \tau}d\omega \\
&= \frac{1}{2\pi}\frac{S_0}{2\gamma m^2}\int_{-\infty}^{\infty}\left[\frac{1}{i\omega + \gamma} + \frac{1}{-i\omega + \gamma}\right]e^{i\omega \tau}d\omega \\
&= \frac{1}{2\pi}\frac{S_0}{2\gamma m^2}\int_{-\infty}^{\infty}\frac{e^{i\omega \tau}}{i\omega + \gamma}d\omega = \frac{1}{2\pi}\frac{S_0}{2\gamma m^2}e^{-\gamma \tau}.
\end{aligned} \tag{4.104}$$

Thus, by taking $\tau = 0$, the above equation becomes

$$\langle v(t_0)v(t_0)\rangle = \frac{1}{2\pi}\frac{S_0}{2\gamma m^2} \text{ or } S_0 = 4\pi\gamma m^2 \langle v(t_0)^2\rangle = 4\pi\gamma mk_BT \text{ by the equipartition law.}$$

If the random force has this property, from equation (4.104),

$$\langle v(t_0)v(t_0+\tau)\rangle = \frac{1}{2\pi}\frac{4\pi\gamma mk_BT}{2\gamma m^2}e^{-\gamma\tau} = \frac{k_BT}{m}e^{-\gamma\tau}. \qquad (4.105)$$

This equation is equivalent to equation (4.92).

### 4.5.4 Random frequency modulation and motional narrowing

Imagine that a harmonic oscillation is being modulated randomly. When we measure its oscillation spectrum, how is the spectrum line-shape dependent on the random modulation? Let us find this out. The equation of the harmonic oscillation is $dx(t)/dt = i\omega(t)x(t)$. Suppose $\omega(t) = \omega_0 + \omega_1(t)$ where $\omega_1(t)$ is assumed to be a randomly modulated variable. The solution of the harmonic oscillation is given by

$$x(t) = x_0 \exp\left[i\int_0^t \omega(\tau)d\tau\right] = x_0 \exp\left[i\int_0^t \{\omega_0 + \omega_1(\tau)\}d\tau\right]$$
$$= x_0 e^{i\omega_0} \exp\left[i\int_0^t \omega_1(\tau)d\tau\right] \qquad (4.106)$$

where we consider $x(t)$ is complex.

According to the Winer–Khintchine theorem (appendix A7), the power spectrum density of the oscillation is given by the correlation function. For a stationary random process, we can consider the following correlation function where we used equation (4.106).

$$\left\langle x^*(0)x(t)\right\rangle = \left\langle x_0^*x_0 \exp\left[i\int_0^t \omega(\tau)d\tau\right]\right\rangle = |x_0|^2 e^{i\omega_0 t}\left\langle\exp\left[i\int_0^t \omega_1(\tau)d\tau\right]\right\rangle. \qquad (4.107)$$

Using equation (4.107), define

$$C(t) \equiv \frac{\langle x^*(0)x(t)\rangle}{|x_0|^2} = e^{i\omega_0 t}\phi(t) \qquad (4.108)$$

where

$$\phi(t) = \left\langle\exp\left[i\int_0^t \omega_1(\tau)d\tau\right]\right\rangle. \qquad (4.109)$$

Define the essential part of the power spectrum of the oscillation, $S(\omega_1)$, using the correlation function (4.109):

$$S(\omega_1) = \int_{-\infty}^{\infty} \phi(t)e^{-i\omega_1 t}dt. \qquad (4.110)$$

Assuming that for a stationary Gaussian modulation with a mean zero, and a correlation function of the following form:

$$\langle \omega_1(t) \rangle = 0$$
$$\langle \omega_1(0)\omega_1(t) \rangle = \langle \omega_1^2 \rangle \psi(t). \tag{4.111}$$

As shown in appendix A8, we obtain

$$\phi(t) = \left\langle \exp\left[ i \int_0^t \omega_1(\tau)d\tau \right] \right\rangle = \exp\left[ -\langle \omega_1^2 \rangle \int_0^t (t - \tau)\psi(\tau)d\tau \right] \tag{4.112}$$

and the power spectrum (4.110) becomes

$$S(\omega_1) = \int_{-\infty}^{\infty} \exp\left[ -i\omega_1 t - \langle \omega_1^2 \rangle \int_0^t (t - \tau)\psi(\tau)d\tau \right]. \tag{4.113}$$

Suppose $\psi(t)$ in the correlation function (4.112) has an exponential decay given by $\psi(t) = \exp(-t/\tau_c)$ where $\tau_c$ represent the decay time of the correction. Equation (4.112) can be computed to become

$$\phi(t) = \exp[-\langle \omega_1^2 \rangle \tau_c[t - \tau_c(1 - e^{-t/\tau_c})]]$$
$$= \exp\left[ -\alpha^2 \left( \frac{t}{\tau_c} - 1 + e^{-t/\tau_c} \right) \right] \tag{4.114}$$

where $\alpha = \langle \omega_1^2 \rangle \tau_c$. As $\tau_c \to \infty$ (or $t \ll \tau_c$), $\alpha \to \infty$, and to the second order of $t$, we obtain

$$\phi(t) \approx \exp\left[ -\alpha^2 \left( \frac{t}{\tau_c} - 1 + \left( 1 - \frac{t}{\tau_c} + \frac{t^2}{2\tau_c^2} \right) \right) \right] = \exp\left[ -\frac{1}{2}\frac{\alpha^2}{\tau_c^2}t^2 \right]$$
$$= \exp\left[ -\frac{1}{2}\langle \omega_1^2 \rangle t^2 \right]. \tag{4.115}$$

In such cases, using appendix A9, the power spectrum becomes Gaussian:

$$\frac{1}{2\pi}S(\omega_1) = \frac{1}{2\pi} \int_{-\infty}^{\infty} \phi(t)e^{-i\omega_1 t}dt = \frac{1}{\sqrt{2\pi\langle \omega_1^2 \rangle}} \exp\left( -\frac{1}{2}\frac{\omega_1^2}{\langle \omega_1^2 \rangle} \right). \tag{4.116}$$

Therefore, as the random modulation rate is rapid, $\langle \omega_1^2 \rangle$ is smaller when the spectrum (4.116) becomes narrower. This phenomenon is called motional narrowing, and may be observed in phenomena of random frequency modulations. In particular, motional narrowing is observed in free induction decay (FID) signals of NMR introduced in section 4.3.2. In NMR, equation (4.115) represents the FID because it is the Fourier transform of the frequency spectrum [15].

As an example, figure 4.41 shows FIDs of solid hydrogen of low concentration of ortho-$H_2$ (about 1%) at absolute temperatures of 4.2 K, 1 K, and 0.05 K. At 4.2 K, a nuclear spin motion is 'rapid' and the frequency spectrum is narrow or the FID is

**Figure 4.41.** NMR signal (FID) of solid hydrogen of low ortho-$H_2$ concentration at 4.2 K, 1.06K, and 0.0 5K. Note: the initial ringing pattern due to the intense RF pulse is eliminated in these figures. Figure courtesy of Don Fujito.

broad. Upon cooling the sample, the motion of nuclear spin becomes slower to cause a larger fluctuation of the local field, and the spectrum becomes broader or the FID becomes narrower, indicting the greater local field effect. The FID at 0.05 K has a very short time duration and side bands appear due to the large fluctuation of the local field caused by spin interactions [16].

## 4.6 Solving diffusion equation using Fourier transform

### 4.6.1 Diffusion equation

Heat equation (1.4) belongs to a category of diffusion equations that describe flows of mass and thermal energies. Consider the following three-dimensional diffusion equation

$$\frac{\partial u(\vec{r}, t)}{\partial t} = D\left(\frac{\partial u(\vec{r}, t)}{\partial x^2} + \frac{\partial u(\vec{r}, t)}{\partial y^2}\frac{\partial u(\vec{r}, t)}{\partial z^2}\right) \tag{4.117}$$

where $D$ is the diffusion coefficient. In this section, rather than the variable separation method, we apply the Fourier transform of $u(\vec{r}, t)$ and the inverse Fourier transform [17].

$$\begin{cases} FT[u(\vec{r}, t)] = \dfrac{1}{\sqrt{8\pi^3}} \displaystyle\int_{-\infty}^{+\infty} u(\vec{r}, t)e^{-i\vec{k}\cdot\vec{r}}dx \equiv U(\vec{k}, t) \\[3mm] FT^{-1}[U(\vec{k}, t)] = \dfrac{1}{\sqrt{8\pi^3}} \displaystyle\int_{-\infty}^{+\infty} U(\vec{k}, t)e^{+i\vec{k}\cdot\vec{r}}dk = u(\vec{r}, t). \end{cases} \tag{4.118}$$

By replacing $u(\vec{r}, t)$ with $FT^{-1}[U(\vec{k}, t)]$, the diffusion equation (4.117) becomes

$$\frac{\partial}{\partial t}\left[\frac{1}{\sqrt{8\pi^3}}\int_{-\infty}^{+\infty} U(\vec{k}, t)e^{+i\vec{k}\cdot\vec{r}}dk\right] = D\left(\frac{\partial^2}{\partial x^2} + \frac{\partial^2}{\partial y^2} + \frac{\partial^2}{\partial z^2}\right)\left[\frac{1}{\sqrt{8\pi^3}}\int_{-\infty}^{+\infty} U(\vec{k}, t)e^{+i\vec{k}\cdot\vec{r}}dk\right].$$

The position derivatives in the above equation can be computed, the equation then becomes

$$\frac{1}{\sqrt{8\pi^3}}\left[\int_{-\infty}^{+\infty}\frac{\partial U(\vec{k}, t)}{\partial t}e^{+i\vec{k}\cdot\vec{r}}dk\right] = -\frac{1}{\sqrt{8\pi^3}}\left[\int_{-\infty}^{+\infty} D(k_x^2 + k_y^2 + k_z^2)U(\vec{k}, t)e^{+i\vec{k}\cdot\vec{r}}dk\right]. \tag{4.119}$$

Thus,

$$\frac{\partial U(\vec{k}, t)}{\partial t} = -D(k_x^2 + k_y^2 + k_z^2)U(\vec{k}, t), \tag{4.120}$$

and its solution is given by

$$U(\vec{k}, t) = U(\vec{k}, 0)e^{-D(k_x^2 + k_y^2 + k_z^2)t}, \tag{4.121}$$

including the initial condition $U(\vec{k}, 0)$, which is defined as

$$U(\vec{k}, 0) = FT[u(\vec{r}, 0)] = \frac{1}{\sqrt{8\pi^3}}\int_{-\infty}^{+\infty} u(\vec{r}, 0)e^{-i\vec{k}\cdot\vec{r}}dx. \tag{4.122}$$

Suppose all particles are at the origin initially, the initial condition can be given by $u(\vec{r}, 0) = \delta(\vec{r}) \equiv \delta(x)\delta(y)\delta(z)$, then equation (4.122) can be computed to get

$$U(\vec{k}, 0) = \frac{1}{\sqrt{8\pi^3}} \left( \int_{-\infty}^{+\infty} \delta(x)e^{-ik_x x} dk_x \right) \left( \int_{-\infty}^{+\infty} \delta(y)e^{-ik_y y} dk_y \right) \left( \int_{-\infty}^{+\infty} \delta(x)e^{-ik_z z} dk_z \right) = \frac{1}{\sqrt{8\pi^3}},$$

and we obtain

$$U(\vec{k}, t) = U(\vec{k}, 0)e^{-D(k_x^2+k_y^2+k_z^2)t} = \frac{1}{\sqrt{8\pi^3}}e^{-D(k_x^2+k_y^2+k_z^2)t}. \tag{4.123}$$

Therefore, the solution, $u(\vec{r}, t)$, with the initial condition $u(\vec{r}, 0) = \delta(\vec{r})$ is given by

$$u(\vec{r}, t) = \frac{1}{\sqrt{8\pi^3}} \int_{-\infty}^{+\infty} U(\vec{k}, t)e^{+i\vec{k}\cdot\vec{r}} d\vec{k} = \frac{1}{8\pi^3} \int_{-\infty}^{+\infty} e^{-D(k_x^2+k_y^2+k_z^2)t} e^{+i\vec{k}\cdot\vec{r}} d\vec{k}$$

$$= \frac{1}{8\pi^3} \left[ e^{-(x^2/4Dt)} \int_{-\infty}^{+\infty} dk_x e^{-Dt[k_x-(ix/2Dt)^2]} \right] \left[ e^{-(y^2/4Dt)} \int_{-\infty}^{+\infty} dk_y e^{-Dt[k_y-(iy/2Dt)^2]} \right] \tag{4.124}$$

$$\times \left[ e^{-(z^2/4Dt)} \int_{-\infty}^{+\infty} dk_z e^{-Dt[k_z-(iz/2Dt)^2]} \right] = \frac{1}{\sqrt{(4\pi Dt)^3}} e^{-[(x^2+y^2+z^2)/4Dt]}.$$

This three-dimensional Gaussian distribution shows the time dependence of diffusing particles, which are assumed to be initially concentrated at the origin. The distribution spreads out as the standard deviation of the Gaussian distribution function (4.124) indicates that $\langle x(t)^2 + y(t)^2 + z(t)^2 \rangle$ is proportional to $Dt$, which is equivalent to Einstein's relation (4.95).

### 4.6.2 Green's function for wave equation

**Green's function for Helmholtz equation:** the space part of the standing wave equation (1.2) and heat/diffusion equation (1.18) belong to the Helmholtz equation. If an external driving force is added to the equation, it becomes an inhomogeneous equation. In the three-dimensional coordinates, the inhomogeneous Helmholtz equation is given by

$$(\nabla^2 + \lambda^2)\psi(\vec{r}) = f(\vec{r}) \tag{4.125}$$

where the term $f(\vec{r})$ may be called the external driving force. How can we solve it? Recall the linear response theory (section 2.4). The solution of equation (4.125) can be expressed with

$$\psi(\vec{r}) = \int f(\vec{r}')G(\vec{r}-\vec{r}')d\vec{r}'. \tag{4.126}$$

The function $G(\vec{r})$ in equation (4.126) is the impulse response, which satisfies

$$(\nabla^2 + \lambda^2)G(\vec{r}) = \delta(\vec{r}) \tag{4.127}$$

where $\delta(\vec{r})$ is the delta function. The solution of the differential equation (4.127) is called Green's function. From the discussion of the linear response theory, Green's function is the impulse response.

We can obtain the Green's function by applying a Fourier transform. Define the Fourier transform of the Green's function as

$$\Gamma(\vec{k}) = \int G(\vec{r})e^{-i\vec{k}\cdot\vec{r}}d\vec{k} \text{ or } G(\vec{r}) = \frac{1}{8\pi^3}\int\Gamma(\vec{k})e^{i\vec{k}\cdot\vec{r}}d\vec{k}. \tag{4.128}$$

Using equations (4.127) and (4.128), we obtain

$$(\nabla^2 + \lambda^2)G(\vec{r}) = \frac{1}{8\pi^3}\int(\lambda^2 - k^2)\Gamma(\vec{k})e^{i\vec{k}\cdot\vec{r}}d\vec{k} = \delta(\vec{r}). \tag{4.129}$$

Recall that the delta function can be expressed as $\delta(\vec{r}) = \frac{1}{8\pi^3}\int e^{i\vec{k}\cdot\vec{r}}d\vec{k}$ from equation (2.19).

Thus, from equation (4.129), we obtain $(\lambda^2 - k^2)\Gamma(k) = 1$, and

$$\Gamma(\vec{k}) = \frac{1}{(\lambda^2 - k^2)} = \int G(\vec{r})e^{-i\vec{k}\cdot\vec{r}}d\vec{k}. \tag{4.130}$$

Here, instead of taking the inverse Fourier transform of $\Gamma(k)$ to find $G(r)$, consider the Fourier transform of $\exp(\pm\lambda r)/r$. Define

$$I = \int\frac{e^{\pm i\lambda r}}{r}e^{-i\vec{k}\cdot\vec{r}}d\vec{r}. \tag{4.131}$$

Using the polar coordinates and referring to appendix A10, we obtain

$$I = \int_0^\infty r^2dr\int_0^\pi \sin\theta d\theta\int_0^{2\pi}d\phi\frac{e^{\pm i\lambda r}}{r}e^{-ikr\cos\theta} = 4\pi\frac{1}{k^2 - \lambda^2}. \tag{4.132}$$

Combining equations (4.130) and (4.131),

$$\Gamma(\vec{k}) = \frac{1}{k^2 - \lambda^2} = \frac{1}{4\pi}I = \frac{1}{4\pi}\int\frac{e^{\pm i\lambda r}}{r}e^{-i\vec{k}\cdot\vec{r}}d\vec{r} = \int G(\vec{r})e^{-i\vec{k}\cdot\vec{r}}d\vec{k}$$

therefore, the Green's function for the differential equation (4.127) is

$$G(r) = \frac{e^{\pm i\lambda r}}{4\pi r} \text{ where } G(r) \to 0 \text{ as } r \to \infty. \tag{4.133}$$

Notice that the Green's function appears as the Huygens' principle in optics (section 4.4), and the solution of the potential scattering in quantum mechanics (section 4.7.5).

**Green's function for wave equation:** we can also find the Green's function of the wave equation

$$\left(\frac{\partial^2}{\partial t^2} - v^2\nabla^2\right)u(\vec{r}, t) = f(\vec{r}, t). \tag{4.134}$$

The Green's function for the wave equation satisfies

$$\left(\frac{\partial^2}{\partial t^2} - v^2\nabla^2\right)G(\vec{r}, t) = \delta(\vec{r})\delta(t) \tag{4.135}$$

As we pointed out in the remark in section 2.1.1 (equations (2.9) and (2.10)), for equation (4.135), we define FT with respect time as

$$G(\vec{r}, t) = \frac{1}{2\pi} \int_{-\infty}^{\infty} \Psi(\vec{r}, \omega)e^{i\omega t}d\omega, \qquad (4.136)$$

and equation (4.133) becomes

$$-\int_{-\infty}^{\infty} (\omega^2 + v^2 \nabla^2)\Psi(\vec{r}, \omega)e^{i\omega t}d\omega = \int_{-\infty}^{\infty} \delta(\vec{r})e^{i\omega t}d\omega. \qquad (4.137)$$

Thus, we obtain

$$\left(\nabla^2 + \frac{\omega^2}{v^2}\right)\Psi(\vec{r}, \omega) = -\frac{1}{v^2}\delta(\vec{r}), \qquad (4.138)$$

equation (4.136) is similar to equation (4.127) with $\lambda = \omega/v$, and thus

$$\Psi(\vec{r}, t) = \frac{1}{4\pi v^2 r} \exp\left(\pm i\frac{\omega}{v}\frac{1}{r}\right). \qquad (4.139)$$

Back to equation (4.133), we obtain

$$G(\vec{r}, t) = \frac{1}{2\pi} \int_{-\infty}^{\infty} \Psi(\vec{r}, \omega)e^{-i\omega t}d\omega = \frac{1}{2\pi} \int_{-\infty}^{\infty} \frac{1}{4\pi v^2 r} \exp\left(\pm i\frac{\omega}{v}r - i\omega t\right)d\omega$$
$$= \frac{1}{2\pi}\frac{1}{4\pi v^2 r} \int_{-\infty}^{\infty} \exp\left[-i\omega\left(t \mp \frac{r}{v}\right)\right]d\omega = \frac{1}{4\pi v^2 r}\delta\left(t \mp \frac{r}{v}\right),$$
$$\qquad (4.140)$$

and the solution of the wave equation (4.144) is given by

$$u(\vec{r}, t) = \iint f(\vec{r}', t')G(\vec{r} - \vec{r}', t - t')d\vec{r}'dt'. \qquad (4.141)$$

For more Green's functions for other differential equations, refer to [18].

## 4.7 Quantum mechanics

### 4.7.1 Wave packet

In quantum mechanics, a particle is treated as a wave. More precisely, a particle is expressed as a localized wave or a wave packet in space. Let us investigate a one-dimensional wave packet, which leads us to two important concepts: group velocity and the uncertainty principle [19].

An arbitrary localized wave propagating in the $x$-direction can be created by superposing plane waves $F(k)e^{i(kx-\omega t)}$ of amplitude $F(k)$:

$$f(x, t) = \int_{-\infty}^{+\infty} F(k)e^{i(kx-\omega t)}dk \qquad (4.142)$$

where the 'momentum amplitude' $F(k)$ is

**Figure 4.42.** Wave packet.

$$F(k) = \frac{1}{2\pi} \int_{-\infty}^{+\infty} f(x, t)e^{-ikx}dk. \tag{4.143}$$

A particle described by equation (4.142) has a distribution of momentum, i.e. uncertainty of momentum. Figure 4.42 shows a wave packet $f(x, 0)$ where momentum has a truncated uniform distribution. This is analogous to the single slit diffraction depicted by figures 4.27 and 4.28.

### 4.7.2 Group velocity

Does the phase speed of the material wave given by $v_p = \lambda{\cdot}f = \omega/k$ correspond to the particle velocity $v = p/m$? From de Broglie's hypothesis, the wavelength is expressed as $\lambda = h/p$, where $p$ is the momentum of a particle. If the particle has the frequency $f$, its energy is given by $E = hf$. Thus, using the relativistic expressions of energy and momentum, the phase velocity is given by

$$v_{\text{wave}} = f \cdot \lambda = \frac{\omega}{k} = \frac{E}{p} = \frac{mc^2}{mv_{\text{particle}}} = \frac{c^2}{v_{\text{particle}}}. \tag{4.144}$$

Therefore, from equation (4.144), $v_p > c$ would be allowed because $v_{\text{particle}} < c$. This is not acceptable and we need to apply the group velocity $v_{\text{group}} = d\omega/dk$, which was originally introduced as the velocity of an electrical signal.

In terms of the particle's parameter,

$$v_{\text{group}} = \frac{d\omega}{dk} = \frac{d(E/\hbar)}{d(p/\hbar)} = \frac{dE}{dp}. \tag{4.145}$$

For a particle, $E = p^2/2m$, and thus $v_{\text{group}} = p/m = v$. This argument can be generalized with the concept of a Fourier transform. If the distribution $F(k)$ of

equation (4.141) is localized around some value $k_0$, then the angular frequency $\omega(k)$ can be expanded around $k_0$:

$$\omega(k) = \omega_0 + \frac{d\omega}{dk}|_0(k - k_0) + \dots \tag{4.146}$$

Thus, we obtain

$$f(x, t) = \int_{-\infty}^{+\infty} F(k)\exp\left\{i\left[kx - \left(\omega_0 t + \frac{d\omega}{dk}\Big|_0 (k - k_0)t + \cdots\right)\right]\right\}dk$$

$$= \exp\left(i\left[\omega_0 t + \frac{d\omega}{dk}\Big|_0 k_0 t\right]\right) + \int_{-\infty}^{+\infty} F(k)\exp\left(i\left[kx - \frac{d\omega}{dk}\Big|_0 kt\right]\right)dk. \tag{4.147}$$

Recalling the definition of a Fourier transform

$$f(x, 0) = \int_{-\infty}^{+\infty} F(k)e^{ikx}dk, \tag{4.148}$$

we obtain

$$\int_{-\infty}^{+\infty} F(k)\exp\left(ik\left[x - \frac{d\omega}{dk}\Big|_0 t\right]\right)dk = f(x', 0) \tag{4.149}$$

where $x' = x - \frac{d\omega}{dk}|_0 t$. Because $d\omega/dk$ is the group velocity $v_g$, the position of the wave packet at a later time $t$ is given by $x' = x - v_g t$. This is equivalent to the position of a particle moving at velocity $v_g$.

### 4.7.3 Schrödinger equation with imaginary time

We focus on the one-dimensional potential problem to introduce the Schrödinger equation with imaginary time. The time-independent Schrödinger equation is given by

$$i\hbar\frac{\partial\psi(x, t)}{\partial t} = \hat{H}\psi(x, t) \text{ where } \hat{H} = -\frac{\hbar^2}{2m}\nabla^2 + V(x). \tag{4.150}$$

For a real-value potential $V(x)$, the wave function $\psi(x)$ is a real function, and we write the time-independent Schrödinger equation as

$$\left[-\frac{\hbar^2}{2m}\frac{d^2}{dx^2} + V(x)\right]\psi(x) = E\psi(x) \text{ where } \int\psi^2 dx = 1. \tag{4.151}$$

Now, back to the time-dependent Schrödinger equation (4.150), if we replace the time variable $t$ with the imaginary time, $\tau = i\hbar t$, the Schrödinger equation becomes

$$\frac{\partial\psi(x, \tau)}{\partial\tau} = -\hat{H}\psi(x, \tau). \tag{4.152}$$

Equation (4.152) can be regarded as a 'diffusion equation' with the imaginary time $\tau$, and we can apply the analytical and numerical methods for solving the classical diffusion equation.

For example, the discussion in section 1.2.2 is applicable to a particle in a one-dimensional potential well:

(1) initial condition: $\psi(x, 0) = f(x)$ for $0 < x < L$, and $u(0,0) = \psi(L, 0) = 0$
(2) boundary condition: $\psi(0, \tau) = \psi(L, \tau) = 0$.

By replacing $\kappa$ with $\hbar^2/2m$ in equation (1.22), the wave function is given by

$$\psi(x, t) = \frac{2}{L}\sum_n c_n e^{-E_n\tau} \sin\left(\frac{n\pi}{L}x\right) \tag{4.153}$$

where $E_n = \frac{\hbar^2}{2m}(\frac{n\pi}{L})^2$. There is an interesting computational approach to solve the Schrödinger equation using the imaginary time diffusion equation. The dominant term of equation (4.153) will be the ground state after a sufficiently long time:

$$\psi(x, t) = \frac{2}{L}c_1 \sin\left(\frac{\pi}{L}x\right) \text{ as } \tau \to \infty. \tag{4.154}$$

Therefore, starting with a best-estimated wave function that satisfies the boundary condition, such a wave function will approach the ground state wave function [20].

### 4.7.4 Heisenberg's uncertainty principle

Heisenberg discovered one of the fundamental principles of quantum physics, the uncertainty principle, which relates to the simultaneous observation of two canonical variables, e.g. position and momentum. Let us define uncertainty in any physically observable quantity by the mean square deviation:

$$(\Delta q)^2 \equiv \langle (q - \langle q \rangle)^2 \rangle \tag{4.155}$$

where $\langle q \rangle$ is the mean value of $q$. Let us define the standard deviation of $q$ as

$$\Delta q \equiv \sqrt{(\Delta q)^2} = \sqrt{\langle (q - \langle q \rangle)^2 \rangle}.$$

The uncertainty of the canonical variable $p$ is also defined as

$$\Delta p \equiv \sqrt{(\Delta p)^2} = \sqrt{\langle (p - \langle p \rangle)^2 \rangle}.$$

With these definitions, the uncertainty principle states

$$(\Delta q) \cdot (\Delta p) \geqslant \frac{\hbar}{2}. \tag{4.156}$$

In extreme cases where $\Delta q = 0$ for a delta function, $\Delta p = \infty$, and vice versa. The former is the case where a particle of definite position has no information on its momentum, and the latter is the case where a plane wave of definite momentum has no information on its position.

Because an introductory course in quantum mechanics often does not point out the fundamental relationship between the uncertainty principle and the Fourier transform, we should investigate how the uncertainty principle originates from the Fourier transform, which connects real space and momentum space.

1. **Uncertainty of Gaussian wave packet:** let us discuss the uncertainty principle using a Gaussian wave packet as a solution to the Schrödinger equation in free space. The normalized wave function of a particle that is localized within a spread of $\sigma$ about the origin moving with an average momentum $k_0$ is given by

$$f(x, 0) = \frac{1}{(2\pi\sigma^2)^{1/4}} e^{-\frac{x^2}{4\sigma^2}} e^{ik_0 x} \text{ at } t = 0. \tag{4.157}$$

The corresponding initial probability density is Gaussian:

$$|f(x, 0)|^2 = \frac{1}{\sqrt{2\pi}\sigma} e^{-\frac{x^2}{2\sigma^2}} \tag{4.158}$$

and the initial uncertainty in the position of the particle in the initial state is $\Delta x = \sigma$. The momentum amplitude corresponding to this initial state is given by

$$F(k) = \frac{1}{2\pi} \int_{-\infty}^{+\infty} f(x, 0) e^{-ikx} dk = \frac{1}{(2\pi)^{3/4}\sigma^{1/2}} \int_{-\infty}^{+\infty} e^{-\left(\frac{x^2}{4\sigma^2}\right)} e^{i(k_0-k)x} dx. \tag{4.159}$$

The Fourier transform of a Gaussian function is also Gaussian (appendix A9), and we obtain

$$F(k) = \frac{1}{(2\pi)^{3/4}\sigma^{1/2}} (4\pi)^{1/2}\sigma e^{-(k_0-k)\sigma^2} = \frac{\sqrt{2}\sigma}{(2\pi)^{1/4}} e^{-(k_0-k)\sigma^2} \tag{4.160}$$

and the momentum amplitude is

$$|F(k)|^2 = \frac{2\sigma}{\sqrt{2\pi}} e^{-2(k_0-k)\sigma^2}. \tag{4.161}$$

From the above equation, the spread or the standard deviation $\Delta k = \Delta p/\hbar$ is $1/2\sigma$, and the following Heisenberg's uncertainty relationship can be obtained:

$$(\Delta x)(\Delta p) = (\sigma)(\hbar/2\sigma) = \hbar/2. \tag{4.162}$$

Thus, the Gaussian wave packet gives the minimum uncertainty. Figures 4.43 and 4.44 show $|f(x,0)|^2$ and $|F(k)|^2$ of the Gaussian wave packet ($\langle x \rangle = 0$ and $\sigma = 1$).

2. **Uncertainty relationship between time and energy:** as discussed in chapter 2, finite size sampling causes side lobes in the spectrum. It is very analogous to the uncertainty principle, and we can summarize the argument. Suppose a

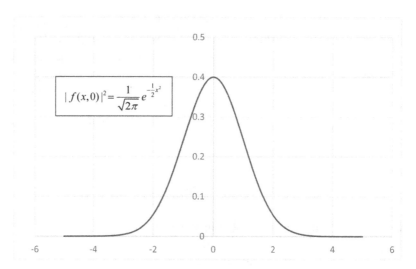

**Figure 4.43.** Gaussian wave packet at $t = 0$.

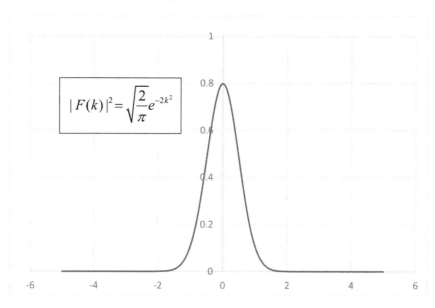

**Figure 4.44.** Gaussian wave packet in $k$-space.

function $y(t)$ is observed in a finite time interval $[-T, + T]$. Its Fourier transform is given by

$$Y_T(\omega) = \int_{-T}^{+T} y(t)e^{-i\omega t}dt \qquad (4.163)$$

whereas the true Fourier transform is given by

$$Y(\omega) = \int_{-\infty}^{+\infty} y(t)e^{-i\omega t}dt. \tag{4.164}$$

Using the square windowing function $w(t) = 1$ for $-T < t < T$; and 0, otherwise, the Fourier transform (4.164) can be written as

$$Y_T(\omega) = \int_{-\infty}^{+\infty} w(t)y(t)e^{-i\omega t}dt. \tag{4.165}$$

By applying the convolution theorem (equation (2.14)), we obtain

$$Y_T(\omega) = \int_{-\infty}^{+\infty} Y(\omega')W(\omega - \omega')d\omega' = \frac{1}{\pi}\int_{-\infty}^{+\infty} Y(\omega')\frac{\sin(\omega'T)}{\omega'}d\omega' \tag{4.166}$$

where we used the Fourier transform of the square window function:

$$W(\omega) = \frac{1}{2\pi}\int_{-\infty}^{+\infty} w(t)e^{-i\omega t}dt = \frac{1}{2\pi}\int_{-T}^{+T} e^{-i\omega t}dt = \frac{1}{\pi}\frac{\sin(\omega T)}{\omega}. \tag{4.167}$$

Similar to figure 4.34, $W(\omega)$ has a maximum at $\omega = 0$ and oscillating and decreasing as $\omega \to \pm\infty$. From equation (4.158), we can interpret that the energy in the frequency domain per unit frequency is given by $|Y(\omega)\frac{\sin(\omega T)}{\omega}|^2$ (excluding some constant coefficient). The distance between the first zeros is $2\pi/T$, and $\pi/T$ can be defined as width $\Gamma$, i.e. $\Gamma = \pi/T$. Thus, we obtain $T/\Gamma = \pi$, which represents an uncertainty relationship between the sampling time and the energy spectrum width.

3. **General wave packet:** as seen, the uncertainty principle is a mathematical consequence of the Fourier transform, and the two variables of the Fourier transform pair satisfy the uncertainty principle. From equation (4.133), a normalized wave packet at a given time can be expressed by

$$f(x, t) = \int_{-\infty}^{+\infty} F(k)e^{i(kx-\omega t)}dk \tag{4.168}$$

where $F(k)$ is the amplitude of a plane wave with wave number $k$, i.e. the wave number or momentum spectrum

$$F(k) = \frac{1}{2\pi}\int_{-\infty}^{+\infty} f(x, t)e^{-ikx}dk. \tag{4.169}$$

Since $|f(x)|^2$ is the probability of finding a particle at $x$, and $|F(k)|^2$ has a similar meaning for the wave number $k$, the mean values of position, $\langle x \rangle$, and the wave number, $\langle k \rangle$ can be defined as

$$\langle x \rangle = \int_{-\infty}^{+\infty} x|f(x)|^2 dx$$
$$\langle k \rangle = \int_{-\infty}^{+\infty} k|F(x)|^2 dk \tag{4.170}$$

and the standard deviations of the uncertainties are given by

$$(\Delta x)^2 = \langle(x - \langle x\rangle)^2\rangle = \int_{-\infty}^{+\infty} (x - \langle x\rangle)^2 |f(x)|^2 dx$$

$$(\Delta k)^2 = \langle(k - \langle k\rangle)^2\rangle = \int_{-\infty}^{+\infty} (k - \langle k\rangle)^2 |F(x)|^2 dk. \tag{4.171}$$

For simplicity, let us consider a wave packet that has $\langle x\rangle = 0$ and $\langle k\rangle = 0$

$$(\Delta x)^2 = \langle x^2\rangle = \int_{-\infty}^{+\infty} x^2 |f(x)|^2 dx$$

$$(\Delta k)^2 = \langle k^2\rangle = \int_{-\infty}^{+\infty} k^2 |F(x)|^2 dk. \tag{4.172}$$

Thus, the following product should be evaluated:

$$(\Delta x)^2 (\Delta k)^2 = \left(\int_{-\infty}^{+\infty} x^2 |f(x)|^2 dx\right) \cdot \left(\int_{-\infty}^{+\infty} k^2 |F(x)|^2 dk\right). \tag{4.173}$$

As shown in appendix A11, the integral $(\Delta x)^2$ can be converted into the following integral

$$(\Delta x)^2 = \int_{-\infty}^{+\infty} \left|\frac{dF(k)}{dk}\right|^2 dk, \tag{4.174}$$

and hence

$$(\Delta x)^2 (\Delta k)^2 = \left(\int_{-\infty}^{+\infty} \left|\frac{dF(k)}{dk}\right|^2 dk\right) \cdot \left(\int_{-\infty}^{+\infty} k^2 |F(x)|^2 dk\right) \geqslant \left(\int_{-\infty}^{+\infty} \left(k|F(k)| \cdot \left|\frac{dF(k)}{dk}\right|\right) dk\right)^2 \tag{4.175}$$

where we applied Schwarz's inequality:

$$\left(\int |\alpha(x)|^2 dx\right) \cdot \left(\int |\beta(x)|^2 dx\right) \geqslant \left(\int |\alpha(x)| \cdot |\beta(x)| dx\right)^2. \tag{4.176}$$

By the partial integral, we obtain

$$\int_{-\infty}^{+\infty} \left(k|F(k)| \cdot \left|\frac{dF(k)}{dk}\right|\right) dk = \frac{1}{2} \int \left(k\frac{d|F(k)|^2}{dk}\right) dk$$

$$= \frac{1}{2}\left[k|F(k)|^2 \Big|_{-\infty}^{+\infty} - \int_{-\infty}^{+\infty} |F(k)|^2 dk\right]. \tag{4.177}$$

Of equation (4.177), the first term is expected to vanish for a localized wave packet, and the second term is one due to the normalization condition. Therefore, equation (4.175) presents Heisenberg's uncertainty principle:

$$(\Delta x)^2 (\Delta k)^2 \geqslant \left(\frac{1}{2}\right)^2 = \frac{1}{4} \text{ or } (\Delta x)^2 (\Delta p)^2 \geqslant \left(\frac{1}{2}\right)^2 = \frac{\hbar^2}{4}. \tag{4.178}$$

In a similar fashion, if $f(t) = \frac{1}{2\pi} \int_{-\infty}^{+\infty} F(\omega)e^{i\omega t}d\omega$, we can prove the uncertainty:

$$(\Delta t)^2(\Delta\omega)^2 \geqslant \frac{1}{4} \text{ or } (\Delta t)^2(\Delta E)^2 \geqslant \frac{\hbar^2}{4} \qquad (4.179)$$

where $F(\omega) = \int_{-\infty}^{+\infty} f(t)e^{-i\omega t}dt$.

### 4.7.5 Potential scattering—Born approximation

It is interesting to note that the Born approximation for a stationary elastic potential scattering uses an analogous approximation of the Fraunhofer diffraction and the x-ray scattering discussed in section 4.5. Let us explain, starting with a Schrödinger equation

$$\left[ -\frac{\hbar^2}{2m}\nabla^2 + V(\vec{r}) \right]\varphi(\vec{r}) = E\varphi(\vec{r}) \qquad (4.180)$$

where $m$ is the reduced mass, and $E = \hbar^2 k^2/2m$ is the kinetic energy of the incident particle. Because the wave function, $\varphi(\vec{r})$, can be written as a superposition of the incident plane wave given by $e^{i\vec{k}\cdot\vec{r}}$ and the scattered wave, $g(\vec{r})$, i.e., $\varphi(\vec{r}) = e^{i\vec{k}\cdot\vec{r}} + g(\vec{r})$, we obtain the following equation for the scattered wave:

$$\left[ -\frac{\hbar^2}{2m}(\nabla^2 + k^2) - V(\vec{r}) \right]g(\vec{r}) = V(\vec{r})e^{i\vec{k}\cdot\vec{r}}. \qquad (4.181)$$

The Born approximation drops the $V(\vec{r})g(\vec{r})$-term as the scattering potential, $V(\vec{r})g(\vec{r})$, is small compared with $V(\vec{r})e^{i\vec{k}\cdot\vec{r}}$. With this approximation, equation (4.181) becomes

$$\left[ -\frac{\hbar^2}{2m}(\nabla^2 + k^2) \right]g(\vec{r}) = V(\vec{r})e^{i\vec{k}\cdot\vec{r}}. \qquad (4.182)$$

The solution of the above equation is given by

$$g(\vec{r}) = -\frac{2m}{\hbar^2}\frac{1}{4\pi} \int \frac{e^{ik|\vec{r}-\vec{r}'|}}{|\vec{r}-\vec{r}'|} V(\vec{r}')e^{i\vec{k}\cdot\vec{r}'} d^3\vec{r}' \qquad (4.183)$$

where vector $\vec{r}'$ shows the potential domain [21].

Analogous to Huygens' principle, equation (4.174) shows that the scattered wave is a superposition of secondary spherical waves with a weight $V(\vec{r}')e^{i\vec{k}\cdot\vec{r}'}$. Notice that the observing position $r$ is much larger than the range of potential $r'$. If $\vec{n}$ is the unit vector in the direction of $\vec{r}$, we obtain

$$|\vec{r}-\vec{r}'| = r - \vec{r}' \cdot \vec{n}, \qquad (4.184)$$

and

$$\frac{e^{ik|\vec{r}-\vec{r}'|}'}{|\vec{r}-\vec{r}'|} = \frac{e^{ik(r-\vec{r}'\cdot\vec{n})}}{r}\left(1 + \frac{\vec{r}'\cdot\vec{n}}{r} + \cdots\right). \tag{4.185}$$

Therefore, denoting $k\vec{n} = \vec{k}'$, the solution (4.177) is given by

$$g(\vec{r}) = -\frac{2m}{\hbar^2}\frac{1}{4\pi}\frac{e^{ikr}}{r}\int V(\vec{r}')e^{i(\vec{k}-\vec{k}')\cdot\vec{r}'}\,d^3\vec{r}' \tag{4.186}$$

to the first order. Similar to the x-ray diffraction (section 4.4.5), we assume the incident and scattered waves are plane waves in the Born approximation.

Let $g(\vec{r}) = \frac{e^{ikr}}{r}f(\theta, \phi)$ where $f(\theta, \phi)$ is called the scattered amplitude, then the Born approximation gives

$$f(\theta, \phi) = -\frac{2m}{\hbar^2}\frac{1}{4\pi}\int V(\vec{r}')e^{i(\vec{k}-\vec{k}')\cdot\vec{r}'}\,d^3\vec{r}' = -\frac{2m}{\hbar^2}\frac{1}{4\pi}\int V(\vec{r}')e^{i\vec{q}\cdot\vec{r}'}\,d^3\vec{r}' \tag{4.187}$$

where $\vec{q} = \vec{k}-\vec{k}'$. Therefore, the Born scattering amplitude is proportional to the spatial Fourier transform of the scattering potential with respect to $\vec{q}$, where $\hbar\vec{q}$ is the transfer of momentum from the incident particle to the scattering potential during the collision. For more discussion, refer to [19, 21].

As we have seen in this chapter, the Fourier transform appears in various physics phenomena, and there are mathematical similarities in the phenomena. It is amazing that the Fourier transform used for acquiring spectra also describes the more fundamental relationship between two spaces of canonical variables.

## References

[1] Horowitz P and Hill W 2015 *The Art of Electronics* (Cambridge: Cambridge University Press)

[2] Hamming R W 1989 *Digital Filters* (New York: Dover)

[3] Cook J D 2016 *Frequency Modulation (FM) and Bessel Functions* (https://johndcook.com/blog/2016/02/17/analyzing-an-fm-signal/)

[4] *FM* Detectors [Analog Devices Wiki]

[5] Bell R 2012 *Introductory Fourier Transform Spectroscopy* (Amsterdam: Elsevier)

[6] Chang R 1971 *Basic Principles of Spectroscopy* (New York: McGraw-Hill) https://amazon.com/Basic-Principles-Spectroscopy-Raymond-Chang/dp/0070105170

[7] Fowlers G R 1989 *Introduction to Modern Optics* (New York: Dover)

[8] Bokor J *Effect of a Lens* (Berkely: University of California) (https://inst.eecs.berkeley.edu›~ee290f›fa04›Lenses_post.pdf)

[9] Kittel C 2005 *Introduction to Solid State Physics* (Internet Archive)

[10] Mahan G D 2011 *Condensed Matter in a Nutshell* (Princeton, NJ: Princeton University Press)

[11] Riley K F, Hobson M P and Bence S J 2002 *Mathematical Methods for Physics and Engineering* (Oxford: Oxford University Press)

[12] Philipse A P 2011 *Notes on Brownian Motion* (https://semanticscholar.org/paper/Notes-on-Brownian-Motion-Philipse/0f300a40087bfa70882d823ab25b61ef2c47a6d4)

[13] Einstein A 1956 *Investigations on The Theory of The Brownian Movement* (Mineola, NY: Dover Publications)

[14] Green W H 2006 *Lecture #31: Fluctuation-Dissipation Theorem* (Boston, MA: MIT)

[15] Abraham A 1978 *The Principles of Nuclear Magnetism* (Oxford: Oxford University Press)

[16] Cho S 1984 *NMR Study of Rotational Fluctuation in Solid Hydrogen and Deuterium* (Columbus, OH: The Ohio State University)

[17] Berntsson F 1999 A spectral method for solving the sideways heat equation *Inverse Prob.* **15** 891 (https://iopscience.iop.org/article/10.1088/0266–5611/15/4/305/meta)

[18] Mathew J and Walker R 1973 *Mathematical Method of Physics* (Menlo Park, CA: Benjamin) Note: there are many lecture notes available from several universities on the Internat

[19] Liboff R L 2003 *Introductory Quantum Mechanics* (Reading, MA: Addison Wesley)

[20] Koonin S E 1987 *Computational Physics* (Menlo Park, CA: Benjamin/Cummings)

[21] Merzbacher E 1998 *Quantum Mechanics* (New York: Wiley)

# Chapter 5

## Quantum Fourier transforms

Fourier transform has been applied to applications of quantum computation. In this chapter, we explain how quantum Fourier transform (QFT) is defined to apply to quantum computation problems such as Shor's algorithm and quantum walks.

### 5.1 Quantum Fourier transform used by Shor's algorithm

The Ron Rivest, Adi Shamir and Leonard Adleman (RSA) cryptosystem uses a prime number that is produced by the multiplication of two large prime numbers because it is extremely difficult to factor large prime numbers. The basic idea of prime factoring is that a prime number has periodicity and if we find the period of a prime number, factoring the prime number can be completed. Although there is no efficient classical algorithm that currently exists for factoring large numbers, Shor demonstrates that it is possible to perform the prime number factoring by using quantum Fourier transform (QFT) based on quantum computation [1]. Below is a brief introduction of qbit and Fourier transform for Shor's algorithm. Because quantum computation uses quantum bits or qbits on which operators are applied to acquire desired quantum states, we will first describe qbit.

#### 5.1.1 Qbit

A qbit represents a two-level quantum state. An electron spin and photon polarization are used as a qbit. When we measure the spin state along the $z$-axis or the 'vertical' axis, which is the default direction of external magnetic field, we observe the spin up ($\uparrow$) or down ($\downarrow$) state. We assign the following vectors $|0\rangle$ and $|1\rangle$ to the spin states:

$$\text{Spin up: } |\uparrow\rangle = |0\rangle = \begin{bmatrix} 1 \\ 0 \end{bmatrix}, \text{ and spin down: } |\downarrow\rangle = |1\rangle = \begin{bmatrix} 0 \\ 1 \end{bmatrix}. \qquad (5.1)$$

doi:10.1088/978-0-7503-6044-9ch5

The spin state vectors form an orthonormal basis of the vector space, $\{|0\rangle, |1\rangle\}$ where $\langle 0|0\rangle = \langle 1|1\rangle = 1$ and $\langle 0|1\rangle = \langle 1|0\rangle = 0$. An arbitrary spin state can be given by $|\psi\rangle = a|0\rangle + b|1\rangle$ where the coefficient $a$ and $b$ are complex value constants. The spin state is normalized, i.e. $|\langle\psi|\psi\rangle|^2 = |a|^2 + |b|^2 = 1$ because either spin state will be observed by a measurement. Mathematical operation of measurement is given by projection operators. The outer products of the unit vectors, $|e_1\rangle\langle e_1|$ and $|e_2\rangle\langle e_2|$, are called the projection operators:

$$\hat{P}_1 = |e_1\rangle\langle e_1| = \begin{bmatrix} 1 \\ 0 \end{bmatrix}[1 \ 0] = \begin{bmatrix} 1 & 0 \\ 0 & 0 \end{bmatrix}, \text{ and } \hat{P}_2 = |e_2\rangle\langle e_2| = [0 \ 1]\begin{bmatrix} 0 \\ 1 \end{bmatrix} = \begin{bmatrix} 0 & 0 \\ 0 & 1 \end{bmatrix} \quad (5.2)$$

where $\hat{P}_1|\psi\rangle = a|0\rangle$, $\hat{P}_2|\psi\rangle = b|1\rangle$, $\hat{P}_1|\psi\rangle + \hat{P}_2|\psi\rangle = |\psi\rangle$, and thus

$$\hat{P}_1 + \hat{P}_2 = \begin{bmatrix} 1 & 0 \\ 0 & 0 \end{bmatrix} + \begin{bmatrix} 0 & 0 \\ 0 & 1 \end{bmatrix} = \begin{bmatrix} 1 & 0 \\ 0 & 1 \end{bmatrix} = \hat{I} \quad (5.3)$$

where $\hat{I}$ is the $2 \times 2$ unit matrix.

For the quantum computation of an $n$-qbit system, $\{|k_{n-1}\rangle|k_{n-2}\rangle \cdots |k_0\rangle\}$, where the total number of data points is $N=2^n$, we denote its quantum state vector as

$$|\vec{x}\rangle = \sum_{k_{n-1}=0}^{1} \sum_{k_{n-2}=0}^{1} \cdots \sum_{k_0=0}^{1} x_{k_{n-1}k_{n-2}\cdots k_0}\{|k_{n-1}\rangle|k_{n-2}\rangle\cdots|k_0\rangle\} \text{ or } |\vec{x}\rangle = \sum_{k=0}^{N-1} x_{\{k\}}|k\rangle. \quad (5.4)$$

The orthonormal basis of the $n$-qbit system is labeled $\{|k\rangle\} = \{|0\rangle, |1\rangle, |2\rangle, \ldots, |N-1\rangle\}$ each of which is a qbit in the *decimal* notation. Notice that we express qbits in the binary format as well as in the decimal format.

**Note:** The *binary number expression of a decimal number $k$* used here is defined as

$$k = k_{n-1}k_{n-2}\cdots k_0 = k_{n-1} \cdot 2^{n-1} + k_{n-2} \cdot 2^{n-2} + \cdots + k_0 \cdot 2^0 \text{ where } k_m = 0 \text{ or } 1, \text{ and } N = 2^n. \quad (5.5)$$

The suffix $\{k\}$ is should be addressed using $N$-numbers: $k_{N-1}\cdots k_1 k_0$ (in binary) or $k$ (decimal) is given by equation (5.4).

With the above notations, a quantum state vector of a single qbit, where $n = 1$ and $N = 2$, is given by

$$\sum_{k=0}^{1} x_k|k\rangle = x_0|0\rangle + x_1|1\rangle \quad (5.6)$$

and a quantum state vector of a 2-qbit system, where $n = 2$ and $N = 4$, is

$$\sum_{k_2=0}^{1} \sum_{k_1=0}^{1} x_{k_2 k_1}|k_2\rangle|k_1\rangle = x_{00}|0\rangle|0\rangle + x_{01}|0\rangle|1\rangle + x_{10}|1\rangle|0\rangle + x_{11}|1\rangle|1\rangle \text{ (in binary)} \quad (5.7)$$
$$= x_0|0\rangle + x_1|1\rangle + x_2|2\rangle + x_3|3\rangle \text{ (in decimal)}.$$

### 5.1.2 Quantum Fourier transform of N-qbit basis

As we described in chapter 2, the discrete *classical* Fourier transform of a $N$-dimensional vector $x = \{x_j; j = 0, 1, 2, ..., N-1\}$ to another $N$-dimensional vector $y = \{y_j; j = 0, 1, 2, ..., N-1\}$ is given by

$$\{x_j\} \xrightarrow[FT]{} \{y_j\} : y_j = \sum_{k=0}^{N-1} x_k e^{i2\pi jk/N}, \tag{5.8}$$

using an orthonormal basis, $\{\exp(i2\pi k/N); k=0, 1, 2, ..., N-1\}$, where the period is $T = 1$.

Now, we define QFT of quantum state vectors by augmenting the classical FT (5.8)

$$|\vec{x}\rangle = \sum_{j=0}^{N-1} x_j |j\rangle \xrightarrow[QFT]{} |\vec{y}\rangle = \sum_{k=0}^{N-1} y_k |k\rangle$$

$$= \sum_{k=0}^{n-1} \left[ \frac{1}{\sqrt{N}} \sum_{j=0}^{N-1} x_k e^{i2\pi jk/N} \right] |k\rangle = \sum_{j=0}^{N-1} x_k \left[ \frac{1}{\sqrt{N}} \sum_{k=0}^{N-1} e^{i2\pi jk/N} |k\rangle \right] \tag{5.9}$$

where $(1/N)^{1/2}$ is added for normalization of qbits.

The above equation can be interpreted as the orthonormal qbit basis $\{|j\rangle\}=\{|j_0\rangle, |j_1\rangle, ..., |j_{N-1}\rangle\}$ (decimal) of $N$-qbit system is Fourier transformed to another orthonormal $N$-qbit basis $\{|k\rangle\}= \{|k_0\rangle, |k_1\rangle, ..., |k_{N-1}\rangle\}$ (decimal):

$$|j\rangle \xrightarrow[QFT]{} \frac{1}{\sqrt{N}} \sum_{k=0}^{N-1} e^{i\frac{2\pi jk}{N}} |k\rangle. \tag{5.10}$$

### Examples of QFT

(1) When $N = 2$, the orthonormal basis is given by $\{|j\rangle\}=\{|0\rangle, |1\rangle\}$, each of which are transformed into:

$$|0\rangle \xrightarrow[QFT]{} \frac{1}{\sqrt{2}} \sum_{k=0}^{1} e^{i2\pi \cdot 0 \cdot k/2} |k\rangle = \frac{1}{\sqrt{2}}(|0\rangle + |1\rangle) \text{ and } |1\rangle \xrightarrow[QFT]{} \frac{1}{\sqrt{2}} \sum_{k=0}^{1} e^{i2\pi \cdot 1 \cdot k/2} |k\rangle$$

$$= \frac{1}{\sqrt{2}}(|0\rangle - |1\rangle). \tag{5.11}$$

This is equivalent to operation of the Hadamard gate (operator):

$$\hat{H} = \frac{1}{\sqrt{2}} \begin{bmatrix} 1 & 1 \\ 1 & -1 \end{bmatrix}; \ \hat{H}|0\rangle = \frac{1}{\sqrt{2}}(|0\rangle + |1\rangle) \text{ and } \hat{H}|1\rangle = \frac{1}{\sqrt{2}}(|0\rangle - |1\rangle). \tag{5.12}$$

**Remark**: It can be shown that

$$\hat{H}|j\rangle = \frac{1}{\sqrt{2}}(|0\rangle + e^{i2\pi 0.j}|1\rangle) \tag{5.13}$$

where $|j\rangle = |0\rangle$ or $|1\rangle$, and '$0.j$' appeared in the exponent of equation (4.11) is the 'binary fraction' defined for the *binary number less than* 1:

$$0.\, j_m j_{m-1} \cdots j_0 = \frac{j_m}{2^1} + \frac{j_{m-2}}{2^2} + \cdots + \frac{j_0}{2^m}. \tag{5.14}$$

Proof: if $j = 0$, $e^{i2\pi 0.j} = e^0 = 1$; and if $j = 1$, $e^{i2\pi 0.j} = e^{i2\pi 0.1} = e^{i\pi} = -1$ as 0.1 (binary) $=1/2$ (decimal).
Thus,

$$|0\rangle + e^{i2\pi 0.j}|1\rangle = |0\rangle + |1\rangle \text{ if } j = 0; \text{ and } |0\rangle + e^{i2\pi 0.j}|1\rangle = |0\rangle - |1\rangle \text{ if } j = 0.$$

Q.E.D.

(2) When $N = 4$, the orthonormal basis of 2-qbit system is given by $\{|0\rangle = |0\rangle_1|0\rangle_0,$ $|1\rangle = |0\rangle_1|1\rangle_0, |2\rangle = |1\rangle_1|0\rangle_0, |3\rangle = |1\rangle_1|1\rangle_0\}$, each of which is transformed into:

$$|0\rangle \xrightarrow[QFT]{} \frac{1}{\sqrt{4}} \sum_{k=0}^{3} e^{i2\pi \cdot 0 \cdot k/4}|k\rangle = \frac{1}{2}(|0\rangle + |1\rangle + |2\rangle + |3\rangle),$$

$$|1\rangle \xrightarrow[QFT]{} \frac{1}{\sqrt{4}} \sum_{k=0}^{3} e^{i2\pi \cdot 1 \cdot k/4}|k\rangle = \frac{1}{2}(|0\rangle + i|1\rangle - |2\rangle - i|3\rangle),$$

$$|2\rangle \xrightarrow[QFT]{} \frac{1}{\sqrt{4}} \sum_{k=0}^{3} e^{i2\pi \cdot 2k \cdot /4}|k\rangle = \frac{1}{2}(|0\rangle - |1\rangle + |2\rangle - |3\rangle),$$

$$|3\rangle \xrightarrow[QFT]{} \frac{1}{\sqrt{4}} \sum_{k=0}^{3} e^{i2\pi \cdot 3 \cdot k/4}|k\rangle = \frac{1}{2}(|0\rangle - i|1\rangle - |2\rangle + i|3\rangle).$$

(3) QFT of $|9\rangle = |1001\rangle$ where $N=2^4=16$ is given by

$$|9\rangle \xrightarrow[QFT]{} \frac{1}{\sqrt{16}} \sum_{k=0}^{15} e^{i\frac{2\pi 9k}{16}}|k\rangle = \frac{1}{4} \sum_{k=0}^{15} e^{i\frac{9k}{8}\pi}|k\rangle$$

$$= \frac{1}{4}(|0\rangle - e^{i\frac{\pi}{8}}|1\rangle + e^{i\frac{\pi}{4}}|2\rangle - e^{i\frac{3\pi}{8}}|3\rangle + e^{i\frac{\pi}{2}}|4\rangle - e^{i\frac{5\pi}{8}}|5\rangle + e^{i\frac{3\pi}{4}}|6\rangle - e^{i\frac{7\pi}{8}}|7\rangle$$

$$- |8\rangle + e^{i\frac{\pi}{8}}|9\rangle - e^{i\frac{\pi}{4}}|10\rangle + e^{i\frac{3\pi}{8}}|11\rangle - e^{i\frac{\pi}{2}}|12\rangle + e^{i\frac{5\pi}{8}}|13\rangle - e^{i\frac{3\pi}{4}}|14\rangle + e^{i\frac{7\pi}{8}}|15\rangle)$$

where $e^{i9\pi/8} = e^{i\pi + i\pi/8} = -e^{i\pi/8}$ etc.

### 5.1.3 Quantum Fourier transform of orthonormal basis

In order to apply QFT in quantum computation, the qbit basis needs to be Fourier transformed which is defined by equation (5.10). For the $n$-qbit system ($N=2^n$), *QFT* becomes

$$|j\rangle \xrightarrow[QFT]{} \frac{1}{\sqrt{2^n}} \sum_{k=0}^{2^n-1} e^{i\frac{2\pi jk}{2^n}}|k\rangle = \frac{1}{\sqrt{2^n}}\left(|0\rangle + e^{i2\pi j\frac{1}{2}}|1\rangle\right)_{n-1} \left(|0\rangle + e^{i2\pi j\frac{1}{2^2}}|1\rangle\right)_{n-2} \cdots \left(|0\rangle + e^{i2\pi j\frac{1}{2^n}}|1\rangle\right)_0 \tag{5.15}$$

where suffixes are the $n$-qbit line (label) numbers from the highest $n-1$ to the lowest 0.

Proof: from equations (5.5) to (5.10) and using the binary format, we obtain

$$|j\rangle \xrightarrow[\text{QFT}]{} \frac{1}{\sqrt{2^n}} \sum_{k=0}^{2^n-1} e^{i\frac{2\pi jk}{2^n}}|k\rangle$$

$$= \frac{1}{\sqrt{2^n}} \sum_{k_{n-1}=0}^{1} \sum_{k_{n-2}=0}^{1} \cdots\cdots \sum_{k_0=0}^{1} e^{i\frac{2\pi j(k_{n-1}2^{n-1}+k_{n-2}2^{n-2}+\cdots\cdots+k_0 2^0)}{2^n}}|k_{n-1}\rangle|k_{n-2}\rangle\cdots\cdots|k_0\rangle$$

$$= \frac{1}{\sqrt{2^n}} \sum_{k_{n-1}=0}^{1} \sum_{k_{n-2}=0}^{1} \cdots\cdots \sum_{k_0=0}^{1} e^{i2\pi j\left(\frac{k_{n-1}}{2}+\frac{k_{n-2}}{2^2}+\cdots\cdots+\frac{k_0}{2^n}\right)}|k_{n-1}\rangle|k_{n-2}\rangle\cdots\cdots|k_0\rangle$$

$$= \frac{1}{\sqrt{2^n}} \sum_{k_{n-1}=0}^{1} \sum_{k_{n-2}=0}^{1} \cdots\cdots \sum_{k_0=0}^{1} \left(e^{i2\pi j\frac{k_1}{2}}|k_{n-1}\rangle\right)_{n-1}\left(e^{i2\pi j\frac{k_2}{2^2}}|k_{n-2}\rangle\right)_{n-2}\cdots\cdots\left(e^{i2\pi j\frac{k_n}{2^n}}|k_0\rangle\right)_0$$

$$= \frac{1}{\sqrt{2^n}}\left(\sum_{k_{n-1}=0}^{1} e^{i2\pi j\frac{k_{n-1}}{2}}|k_1\rangle\right)_{n-1}\left(\sum_{k_{n-2}=0}^{1} e^{i2\pi j\frac{k_{n-2}}{2^2}}|k_{n-2}\rangle\right)_{n-2}\cdots\cdots\left(\sum_{k_n=0}^{1} e^{i2\pi j\frac{k_0}{2^n}}|k_0\rangle\right)_0$$

$$= \frac{1}{\sqrt{2^n}}\left(|0\rangle + e^{i2\pi j\frac{1}{2}}|1\rangle\right)_{n-1}\left(|0\rangle + e^{i2\pi j\frac{1}{2^2}}|1\rangle\right)_{n-2}\cdots\cdots\left(|0\rangle + e^{i2\pi j\frac{1}{2^n}}|1\rangle\right)_0.$$

(5.16)

Q.E.D.

For implementing QFT of equation (5.15), staring with $n$-qbits $\{|0\rangle\}$, we use two quantum gates (operators). They are the Hadamard gate given by equation (5.12) and a rotational gate defined as

$$\hat{R}_{[m]}|0\rangle = |0\rangle \text{ and } \hat{R}_{[m]}|1\rangle = e^{i2\pi/2^m}|1\rangle \text{ to generate } e^{i2\pi/2^m}; \ m = 1, 2, \ldots, n \quad (5.17)$$

where the suffix $[m]$ specifies the rotational angle of $2\pi/2^m$.

For example, QFT of $|9\rangle = |1\rangle_3|0\rangle_2|0\rangle_1|1\rangle_0$, where suffixes are qbit line numbers from the highest (3) to the lowest (0), can be formulated as

$$(\hat{H})_0\left(\hat{R}_{[2]}\hat{H}\right)_1\left(\hat{R}_{[3]}\hat{R}_{[2]}\hat{H}\right)_2\left(\hat{R}_{[4]}\hat{R}_{[3]}\hat{R}_{[2]}\hat{H}\right)_3|1\rangle_3|0\rangle_2|0\rangle_1|1\rangle_0$$

$$= \frac{1}{\sqrt{2}}\left(|0\rangle - e^{i\frac{\pi}{8}}|1\rangle\right)_3\frac{1}{\sqrt{2}}\left(|0\rangle + e^{i\frac{\pi}{4}}|1\rangle\right)_2\frac{1}{\sqrt{2}}\left(|0\rangle + e^{i\frac{\pi}{2}}|1\rangle\right)_1\frac{1}{\sqrt{2}}(|0\rangle - |1\rangle)_0$$

$$\xrightarrow[\text{qbit-swapping}]{} \frac{1}{4}(|0\rangle - |1\rangle)_3\left(|0\rangle + e^{i\frac{\pi}{2}}|1\rangle\right)_2\left(|0\rangle + e^{i\frac{\pi}{4}}|1\rangle\right)_1\left(|0\rangle - e^{i\frac{\pi}{8}}|1\rangle\right)_0$$

$$= \frac{1}{4}(|0000\rangle - e^{i\frac{\pi}{8}}|0001\rangle + e^{i\frac{\pi}{4}}|0010\rangle - e^{i\frac{3\pi}{8}}|0011\rangle + e^{i\frac{\pi}{2}}|0100\rangle - e^{i\frac{5\pi}{8}}|0101\rangle$$

$$+ e^{i\frac{3\pi}{4}}|0110\rangle - e^{i\frac{7\pi}{8}}|0111\rangle - |1000\rangle + e^{i\frac{\pi}{8}}|1001\rangle - e^{i\frac{\pi}{4}}|1010\rangle + e^{i\frac{3\pi}{8}}|1011\rangle$$

$$- e^{i\frac{\pi}{2}}|1100\rangle + e^{i\frac{5\pi}{8}}|1101\rangle - e^{i\frac{3\pi}{4}}|1110\rangle + e^{i\frac{7\pi}{8}}|1111\rangle)$$

$$= QFT|9\rangle$$

(5.18)

where qbit-swapping is required after applying quantum gates in order to complete QFT.

Shor's algorithm for factorizing prime numbers uses the *inverse* quantum Fourier transform $(QFT^{-1})$, which is the backward operation of QFT, applying the *reverse* gates used in QFT. They are $\hat{H}^{-1} = \hat{H}$, $\hat{R}_{[m]}^{-1}|0\rangle = |0\rangle$, and $\hat{R}_{[m]}^{-1}|1\rangle = e^{-i2\pi/2^m}|1\rangle$.

For example, $QFT^{-1}$ of $|9\rangle = |1\rangle_3|0\rangle_2|0\rangle_1|1\rangle_0$ takes the following steps. First, we need to perform qbit-swapping:

$$|9\rangle = |1\rangle_3|0\rangle_2|0\rangle_1|1\rangle_0 \xrightarrow[\text{qbit-swapping}]{} |0\rangle_3|0\rangle_2|1\rangle_1|1\rangle_0.$$

Second, operate the reverse gates:

$$\left(\hat{H}\hat{R}_{[4]}^{-1}\hat{R}_{[3]}^{-1}\hat{R}_{[4]}^{-1}\right)_3\left(\hat{H}\hat{R}_{[3]}^{-1}\hat{R}_{[2]}^{-1}\right)_2\left(\hat{H}\hat{R}_{[2]}^{-1}\right)_1(\hat{H})_0|0\rangle_3|0\rangle_2|1\rangle_1|1\rangle_0$$

$$= \frac{1}{4}(|0\rangle + |1\rangle)_3(|0\rangle - |1\rangle)_2\left(|0\rangle + e^{-i\frac{2\pi}{2^2}}|1\rangle\right)_1\left(|0\rangle + e^{-i\frac{2\pi}{2^3}}|1\rangle\right)_0$$

$$= \frac{1}{4}(|000\rangle + e^{-i\frac{\pi}{4}}|0001\rangle - i|0010\rangle - ie^{-i\frac{\pi}{4}}|0011\rangle - |0100\rangle - e^{-i\frac{\pi}{4}}|0101\rangle$$

$$+ i|0110\rangle + ie^{-i\frac{\pi}{4}}|0111\rangle + |1000\rangle + e^{-i\frac{\pi}{4}}|1001\rangle - i|1010\rangle - ie^{-i\frac{\pi}{4}}|1011\rangle$$

$$- |1100\rangle - e^{-i\frac{\pi}{4}}|1101\rangle + ie^{-i\frac{\pi}{4}}|1111\rangle).$$

(5.19)

Refer to detailed description of how to apply $QFT^{-1}$ to the Shor's algorithm [2, 3].

## 5.2 Quantum Fourier transform used in quantum walks

This section illustrates a one-dimensional quantum walk and how quantum Fourier transform is applied to the quantum walk.

### 5.2.1 Quantum walk

A basic model of a quantum walk can be described in the following manner. Imagine a one-dimensional lattice and assign a qbit on each lattice point at a given time. We define the increments of the lattice point and time to be 1. The qubit at each lattice point, $\{n; n = 0, \pm1, \pm2, \ldots\}$ at time, $t$, can be expressed as

$$|\psi_t(n)\rangle = a_t(n)|0\rangle + b_t(n)|1\rangle = \begin{bmatrix} a_t(n) \\ b_t(n) \end{bmatrix}$$

(5.20)

where $a_t(n)$ and $b_t(n)$ are constants in time. The discrete quantum walk is to trace the time development of qbits under a transition rule described below.

Using qbits at lattice points $n\pm1$ at time $t$, define the qbit at the lattice point $n$ at time $t+1$ as

$$|\psi_{t+1}(n)\rangle = \hat{P}_1\hat{U}|\psi_t(n + 1)\rangle + \hat{P}_2\hat{U}|\psi_t(n - 1)\rangle$$

(5.21)

where $\hat{U}$ is a unitary operator for transition of the qbit. $\hat{P}_1$ and $\hat{P}_2$ are the projection operators (5.3). The time development has two steps, they are (i) transition of the qbit state by the operator $\hat{U}$; and (ii) coordinate shift by $\hat{P}_1$ and $\hat{P}_2$:

$$\hat{U}|\psi_t(n \pm 1)\rangle = \begin{bmatrix} u_{11} & u_{12} \\ u_{21} & u_{22} \end{bmatrix} \begin{bmatrix} a_t(n \pm 1) \\ b_t(n \pm 1) \end{bmatrix} = \begin{bmatrix} u_{11}a_t(x \pm 1) + u_{12}b_t(x \pm 1) \\ u_{21}a_t(x \pm 1) + u_{22}b_t(x \pm 1) \end{bmatrix} \quad (5.22)$$

$$\hat{P}_1\hat{U}|\psi_t(n + 1)\rangle = \begin{bmatrix} u_{11}a_t(n + 1) + u_{12}b_t(n + 1) \\ 0 \end{bmatrix} \text{ and}$$

$$\hat{P}_2\hat{U}|\psi_t(n - 1)\rangle = \begin{bmatrix} 0 \\ u_{21}a_t(n - 1) + u_{22}b_t(n - 1) \end{bmatrix}. \quad (5.23)$$

Thus,

$$|\psi_{t+1}(n)\rangle = \begin{bmatrix} a_{t+1}(n) \\ b_{t+1}(n) \end{bmatrix} = \begin{bmatrix} u_{11}a_t(n + 1) + u_{12}b_t(n + 1) \\ u_{21}a_t(n - 1) + u_{22}b_t(n - 1) \end{bmatrix}. \quad (5.24)$$

If we define

$$\hat{p} = \hat{P}_1\hat{U} \text{ and } \hat{q} = \hat{P}_2\hat{U}, \quad (5.25)$$

$$|\psi_{t+1}(x)\rangle = \hat{p}|\psi_t(x + 1)\rangle + \hat{q}|\psi_t(x - 1)\rangle \text{ where } \hat{p} + \hat{q} = (\hat{P}_1 + \hat{P}_2)\hat{U} = \hat{U} \quad (5.26)$$

figure 5.1 depicts a schematic diagram of the quantum walk. The relationship (5.26) is analogous to a one-dimensional classical random walk where a classical random walker moves to +1 with probability $p$ and −1 with probability $q$, and $p + q = 1$. However, the behavior of the quantum walk after a sufficiently long time is significantly different to that of the classical random walk. A classical walker is described by a binary distribution [4] whereas a quantum walk exhibits a tendency of the probability distribution shown in figure 5.2 [5].

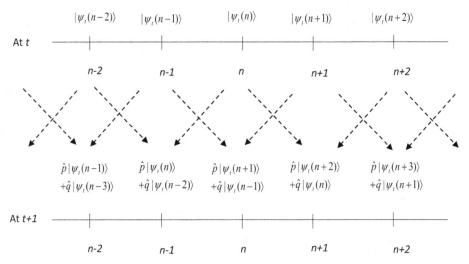

**Figure 5.1.** Time-development of qbits of a quantum walk.

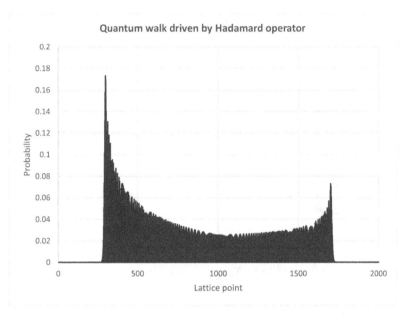

**Figure 5.2.** One-dimensional quantum walk, starting at the center (1001).

**Example of a quantum walk:** figure 5.2 shows the computed quantum walk where the $\hat{U}$ is the Hadamard gate (5.12). The number of lattice points is 2001, indexing from 0 to 2001. The center of the lattice is at 1001 where we start a quantum walk for computational convenience. The number of time steps is 1000. The initial condition is $|\psi_0(m)\rangle = a_0(m)|0\rangle + b_0(m)|1\rangle$ at $m=1001$; and 0 otherwise. Appendix A13.6 lists the VBA code of the quantum walk.

There are two spaces involved in quantum walks: one represented by the lattice points (also called the coordinate space) and the other based on the qbits (also called the coin space) on each lattice point. Combining them together, we describe the quantum walk. The combination is implemented with the direct product of two spaces. In other words, the quantum walk system can be formed by the direct product of two spaces: one with the orthonormal basis $\{|n\rangle; n=0, \pm1, \pm2, \ldots\}$ of the lattice sites, and a two-dimensional qbit-space, $|\psi_t(n)\rangle$ on each lattice site (5.15). The quantum walk is thus described by the following quantum state:

$$|\Psi_t\rangle = \sum_n |n\rangle \otimes |\psi_t(n)\rangle \tag{5.27}$$

and the initial condition is given by $|\psi_0(0)\rangle = a_0(0)|0\rangle + b_0(0)|1\rangle$ and $|\psi_0(x \neq 0)\rangle = 0$.

### 5.2.2 Quantum Fourier transform for the quantum walk

Grimmett *et al* applied quantum Fourier transform to study the asymptotic behavior of the quantum walk after a sufficiently long time [6]. A Fourier transform that is suitable for quantum walks can be defined below. Recalling equation (2.18)

$$\frac{1}{2\pi} \int_{-\pi}^{\pi} e^{ik(n-m)}dk \; = 1 \text{ if } n = m, \text{ and } 0 \text{ otherwise,} \tag{5.28}$$

we can set

$$|\psi_t(n)\rangle = \sum_m \left( \frac{1}{2\pi} \int_{-\pi}^{\pi} e^{ik(n-m)}dk \right)|\psi_t(m)\rangle$$

$$= \cdots + 0 \cdot |\psi_t(n-1)\rangle + 1 \cdot |\psi_t(n)\rangle + 0 \cdot |\psi_t(n+1)\rangle + \cdots. \tag{5.29}$$

Modifying equation (5.29),

$$|\psi_t(n)\rangle = \sum_m \left( \frac{1}{2\pi} \int_{-\pi}^{\pi} e^{ik(n-m)}dk \right)|\psi_t(m)\rangle = \frac{1}{2\pi} \int_{-\pi}^{\pi} e^{ikn}\left( \sum_m e^{-ikm}|\psi_t(m)\rangle \right)dk \tag{5.30}$$

thus, if we define the state of a quantum walker in the $k$-space as

$$|\Psi_t(k)\rangle = \sum_m e^{-ikm}|\psi_t(m)\rangle \tag{5.31}$$

and regarding equation (5.31) as the Fourier transform for the quantum walker. We also define the inverse Fourier transform as

$$|\psi_t(n)\rangle = \frac{1}{2\pi} \int_{-\pi}^{\pi} e^{ikn}|\Psi_t(k)\rangle dk, \tag{5.32}$$

which is the same as equation (5.30).

From equations (5.21) and (5.31), the time evolution of a quantum walker in the $k$-space based on $|\Psi_t(k)\rangle$ is given by

$$|\Psi_{t+1}(k)\rangle = \sum_m e^{-ikm}|\psi_{t+1}(m)\rangle = \sum_m e^{-ikm}\left( \hat{P}_1\hat{U}|\psi_t(m+1)\rangle + \hat{P}_2\hat{U}|\psi_t(m-1)\rangle \right)$$

$$= \sum_m e^{-ik(m-1)}\left( \hat{P}_1\hat{U}|\psi_t(m)\rangle \right) + \sum_m e^{-ik(m+1)}\left( \hat{P}_2\hat{U}|\psi_t(m)\rangle \right)$$

$$= e^{ik}\hat{P}_1\hat{U}\sum_m e^{-ikm}|\psi_t(m)\rangle + e^{-ik}\hat{P}_2\hat{U}\sum_m e^{-ikm}|\psi_t(m)\rangle \tag{5.33}$$

$$= \left( e^{ik}\hat{P}_1 + e^{-ik}\hat{P}_2 \right)\hat{U}\sum_m e^{-ikm}|\psi_t(m)\rangle = \hat{R}(k)\hat{U}\sum_m e^{-ikm}|\psi_t(m)\rangle$$

where $\hat{R}(k) = e^{ik}\hat{P}_1 + e^{-ik}\hat{P}_2 = \begin{bmatrix} e^{ik} & 0 \\ 0 & e^{-ik} \end{bmatrix}$.

Repeatedly applying the operator, $\hat{R}(k)\hat{U}$, to equation (5.29), we obtain

$$|\Psi_{t+1}(k)\rangle = \hat{R}(k)\hat{U}\sum_m e^{-ikm}|\psi_t(m)\rangle = \left( \hat{R}(k)\hat{U} \right)\left( \hat{R}(k)\hat{U} \right)\sum_m e^{-ikm}|\psi_{t-1}(m)\rangle$$

$$= \cdots = \left( \hat{R}(k)\hat{U} \right)^{t+1}\sum_m e^{-ikm}|\psi_0(m)\rangle = \left( \hat{R}(k)\hat{U} \right)^{t+1}|\Psi_0(k)\rangle \tag{5.34}$$

where $\sum_m e^{-ikm}|\psi_0(m)\rangle$ is the initial condition of the walker. For the initial condition, we often take $|\psi_0(m)\rangle = a_0(m)|0\rangle + b_0(m)|1\rangle$ if $m=0$; and 0 otherwise. The corresponding initial condition in the $k$-space is

$$|\Psi_0(k)\rangle = \sum_m e^{-ikm}|\psi_0(m)\rangle = a_0(0)|0\rangle + b_0(0)|1\rangle. \qquad (5.35)$$

The inverse Fourier transform (5.28) provides the time development in the lattice space:

$$|\psi_t(n)\rangle = \frac{1}{2\pi}\int_{-\pi}^{\pi} e^{ikn}|\Psi_t(k)\rangle dk = \frac{1}{2\pi}\int_{-\pi}^{\pi} e^{ikn}(\hat{R}(k)\hat{U})^t|\psi_0(k)\rangle dk. \qquad (5.36)$$

The probability distribution where the quantum walker is observed at position $n$ at time $t$ is given by

$$\langle\psi_t(n)|\psi_t(n)\rangle = |a_t(n)|^2 + |b_t(n)|^2. \qquad (5.37)$$

There are many interesting studies for quantum walk problems. For example, some quantum walks lead to well-known physics partial differential equations, such as the Dirac equation. The time evolution equation of a one-dimensional quantum walker is exactly mapped to the three-dimensional Weyl equation for a zero-mass particle with spin 1/2 [7, 8].

# References

[1] Shor P W 1994 Algorithms for quantum computation: discrete logarithms and factoring *Proc. 35th Annual Symp. on Foundations of Computer Science IEEE Comput. Soc. Press* pp 124–34 (Algorithms for quantum computation: discrete logarithms and factoring | IEEE Conf. Publication | IEEE Xplore)

[2] Cho S 2022 *Quantum Computation and Quantum Information Simulation using Python IOP Series in Quantum Technology* (San Rafael, CA: Morgan & Claypool)

[3] Nielsen M A and Chuang I L 2010 *Quantum Computation and Quantum Information* (Cambridge: Cambridge University Press) (mmrc.amss.cas.cn/tlb/201702/W020170224608149940643.pdf)

[4] Cho S 2023 *Monte Carlo Simulations Using Microsoft EXCEL®* (Chambersburg, PA: Springer)

[5] Nash B 2018 *Introduction to Quantum Walks* (https://windowsontheory.org/2018/12/23/introduction-to-quantum-walks/)

[6] Grimmett G, Jason S and Scudo P F 2004 *Weak limits for quantum random walks Phys. Rev. E* **69** 026119 https://statslab.cam.ac.uk/~grg/papers/sjqrw3.pdf

[7] Arrighi P, Di Molfetta G, Márquez-Martín I and Armando Pérez A 2018 Dirac equations as a quantum walk over the honeycomb and triangular lattices *Phys. Rev. A* **97** 062111 (Phys. Rev. A 97, 062111 (2018)—Dirac equation as a quantum walk over the honeycomb and triangular lattices (aps.org)triangular lattices (aps.org))

[8] Katori M, Fujino S and Konno N 2005 Quantum walks and orbital states of a Weyl particle *Phys. Rev. A* **72** 012316 https://journals.aps.org/pra/abstract/10.1103/PhysRevA.72.012316

**IOP** Publishing

Fourier Transform and Its Applications Using Microsoft EXCEL®
(Second Edition)

**Shinil Cho**

# Chapter 6

## Beyond the Fourier transform spectroscopy

The Fourier transform is not the only method to acquire power spectra. Developed from stochastic processes, the linear prediction (LP) method and the maximum entropy (ME) method can also be applied to estimate power spectra [1]. These methods are actually equivalent for the most probable physical process, the Gaussian process. Although EXCEL does not have built-in routines for these methods, creating an EXCEL macro program of the LP method is straightforward. Furthermore, using the same time-series data, we can compare a power spectral density from the LP method with one from the Fourier transform.

### 6.1 LP method

Let us define the entropy of the $N$ time-series data points of the equal sampling time interval, $\Delta t$, $\{x(k); k = 0, \Delta t, 2\Delta t, \ldots, N\Delta t\}$ for a stochastic process. According to the Winer–Khintchine theorem (appendix A7), the Fourier transform of power spectral density is an autocorrelation function, and the inverse Fourier transform of an autocorrelation function is power spectral density. Therefore, if there is a method of estimating the autocorrelation function for time-series data, the power spectrum density should be obtained by applying the Winer–Khintchine theorem.

The auto-regression (AR) model, applied to time-series data $\{x(k)\}$, predicts the data points at a given time by using data points of previous times:

$$x_{\text{predicted}}(k) = -\sum_{n=1}^{M} d_n x(k - n) \tag{6.1}$$

where $\{d_n\}$ is called the LP coefficient and $M$ is the order of terms used for the estimation. The negative sign is added for computational convenience.

The LP coefficients are determined by minimizing the mean square error:

$$E_p = \frac{1}{N}\sum_{k=0}^{N-1}[w(k)]^2 = \frac{1}{N}\sum_{k=0}^{N-1}\left[x(k) + \sum_{n=1}^{M}d_n x(k-n)\right]^2 \tag{6.2}$$

$$\frac{\partial E_p}{\partial d_j} = 0 \, (1 \leqslant j \leqslant M)$$

where $w(k)$ is an error defined by

$$w(k) \equiv x(k) - x_{\text{predicted}}(k) = x(k) + \sum_{n=1}^{M}d_n x(k-n). \tag{6.3}$$

By performing the minimization (6.2), we obtain

$$x(k)x(k-j) + \sum_{n=1}^{M}d_n x(k-n)x(k-j) = 0 \quad (1 \leqslant j \leqslant M) \tag{6.4}$$

and the expectation value of the minimization condition becomes

$$\langle x(k)x(k-j)\rangle + \sum_{n=1}^{M}d_n\langle x(k-n)x(k-j)\rangle = 0. \tag{6.5}$$

The physical meaning of equation (6.5) can be understood in the following way. Consider the expectation value of $w(k)x(k-j)$ that can be obtained from equation (6.3):

$$\langle w(k)x(k-j)\rangle = \langle x(k)x(k-j)\rangle + \sum_{n=1}^{M}d_n\langle x(k-n)x(k-j)\rangle. \tag{6.6}$$

If the distribution of the error function, $\{w(k)\}$, is Gaussian 'noise', i.e. if $\{w(k)\}$ is Gaussian and independent of $\{x(k)\}$, it is characterized by the zero-mean value and the standard deviation given by

$$\langle w(k)x(k-j)\rangle = \sigma_M^2 \text{ if } j = 0 \tag{6.7}$$
$$= 0 \text{ if } j \neq 0(1 \leqslant j \leqslant M).$$

From equations (6.6) and (6.7), the autocorrelation function, $R(k) = \langle x(k)x(k-j)\rangle$, satisfies the following equation called the Yule–Walker equation [2]

$$R(k) + \sum_{n=1}^{M}d_n R(k-n) = \sigma_M^2 \quad \text{if } k = 0 \tag{6.8}$$
$$= 0 \quad \text{if } 1 \leqslant k \leqslant M.$$

The Levinson algorithm (appendix A13.7) can solve the Yule–Walker equation efficiently to compute a set of the LP coefficients and the standard deviation, $\{d_1, d_2, \ldots, d_M, \sigma_M^2\}$ for a given order $M$. Power spectrum density is then given by

$$S(\omega) = \frac{\sigma_M^2}{\left| 1 + \displaystyle\sum_{n=1}^{M} d_n e^{-i\omega_0 n \Delta t} \right|^2} \tag{6.9}$$

where $\omega_0 = 2\pi f_s/N$ and $f_s = 1/\Delta t$ is the sampling frequency.

EXCEL has no built-in program for the Levinson algorithm, but it is not difficult to create one using EXCEL's VBA macro capability. In order to enable the macro programming, there is a one-time setting. Refer to appendix A12.3 for the enabling of EXCEL's macro and the source program. Appendix A13.8 lists the source code we created for calculating the LP coefficients.

### 6.1.1 Final prediction error

How can the order $M$ be determined properly? For determining the optimal $M$, Akaike suggests final prediction error (FPE) [3] given by the following equation:

$$\text{FPE} = \left( \frac{N + M + 1}{N - M - 1} \right) \delta_M \tag{6.10}$$

where $\delta_M = \displaystyle\sum_{k=M+1}^{N} (w(k))^2$. Generally, the number of terms used for estimation or the order $M$ is optimal when FPE takes the minimum value. Because the Fourier transform-based spectrum for a signal to which the LP analysis is applied is available, we should compare both spectra as the LP-based spectrum closest to the FT-based spectrum may not exactly give the minimum FPE. With EXCEL, we can easily compare them. We demonstrate how the estimated spectrum changes as we change $M$, by comparing with the FT-based spectrum so that readers may obtain the difference visually. The source code of the LP method listed in appendix A13.9 also computes and displays the FPE value.

## 6.2 ME method

Let us define entropy of the $N$ time-sequence data points of an equal sampling time interval $\Delta t$, $\{x(k); k = 0, \Delta t, 2\Delta t, \ldots, N\Delta t\}$, for a stochastic process as

$$H = -\int p(x_0, x_1, x_2, \ldots, x_N) \log p(x_0, x_1, x_2, \ldots, x_N) dx_0 dx_1 dx_2 \cdots dx_N \tag{6.11}$$

where $x_j$ is the value of the $j$th data point $x(j)$, and $p(x_0, x_1, x_2, \ldots, x_N)$ is the joint probability density function. The physically realizable joint probability density function is expected to maximize the entropy [4]. Because of the central limit theorem, the probability density function is most likely Gaussian, which indeed maximizes the entropy.

Assuming that the joint probability density function $p(x_0, x_1, x_2, \ldots, x_N)$ is Gaussian, we define the 'entropy-rate' by the following equation

$$H' = \lim_{N \to \infty} \frac{H}{N+1} = \frac{1}{2}\log(2\pi e) + \frac{1}{4\omega_N}\int_{-\omega_N}^{+\omega_N} \log S(\omega)d\omega \qquad (6.12)$$

where $\omega_N$ is the Nyquist frequency and $S(\omega)$ is the power spectral density. For estimating the spectral density $S(\omega)$, the ME method maximizes the essential part of the above equation, called the time-series entropy,

$$H = \int_{-\omega_N}^{+\omega_N} \log S(\omega)d\omega \qquad (6.13)$$

with a condition of the Winer–Khintchine theorem. This maximization using the Lagrange multiplier method leads to the Yule–Walker equation from the AR method [5].

Suppose the time-series process $\{x(t)\}$ is Gaussian, characterized with the mean $\langle x(t) \rangle$ and the covariance $\langle (x(k)-\langle x(k)\rangle)\cdot(x(m)-\langle x(m)\rangle)\rangle$. The covariance is also called the autocorrelation function. It is also known that a Gaussian process that minimizes the mean square error $\langle (x(k) - \langle x(k)\rangle)^2 \rangle$ is an AR process. Therefore, the ME method is equivalent to the AR process from which we can calculate the LP coefficients, and power spectral density. Mathematically, the undetermined multipliers introduced to the Lagrange multiplier correspond to the liner prediction coefficients.

## 6.3 LPC examples

**Example 1:** triple signal interferogram. Figure 6.1 shows superposed sound signals from three different signal generators whose frequencies are set to 950, 1050, and 1100 Hz, with different signal amplitudes. For calculating the power spectrum, we

**Figure 6.1.** Three sounds of different frequencies and amplitudes.

**Power spectrum by Fourier transform**

Power spectrum of three sinusoidal signals of frequencies 950 Hz, 1050 Hz, and 1100 Hz with different amplitudes.

Frequency (Hz)

**Figure 6.2.** Power spectrum by Fourier transform.

used 4096 terms and applied the Hanning window. Figure 6.2 shows a power spectrum generated by EXCEL's FFT.

Figure 6.3 shows power spectra by the LPC method where we change the order $M$ from 100 to 250. When $M = 180$ to 190, the FPE values reaches near its minimum, and when $M = 185$, the LP-based power spectrum is the closest to the one obtained by using the FT-based spectrum (figure 6.4).

**Example 2:** power spectrum of voice sample.

Figure 6.5 shows an optimal spectrum of the female voice acquired by the LPC analysis. The optimal order $M$ is 380, where the FPE value is minimum and the peak intensity ratios are close to the FT-based spectrum.

The decibel scale spectra obtained by the FFT and LPC are superposed in figure 6.6. LPC correctly reproduces the peak frequencies and overall base line.

## 6.4 LPC cepstrum

In sections 2.5 and 3.7.5, we pointed out that the cepstrum, defined as the inverse Fourier transform of the logarithmic spectrum, can separate the major spectral peaks (formants) of the human voice, which correspond to the resonant frequencies with the vocal tract structure that are superposed over the harmonics.

$$c_y(m) = \frac{1}{N} \sum_{k=0}^{N-1} \log |Y(k)| e^{i2\pi \frac{mk}{N}}. \tag{6.14}$$

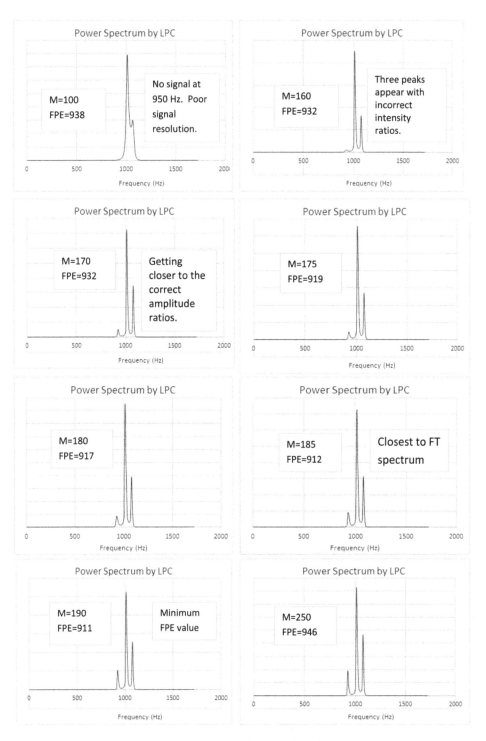

**Figure 6.3.** Power spectrum obtained by LP method ($N$=1024).

**Figure 6.4.** Power spectrum of a female voice by FFT ($N = 4096$).

**Figure 6.5.** Power spectrum constructed by FPC ($N = 4096$ and $M = 380$).

We can also obtain the cepstrum coefficients from the LP coefficients and the standard deviation $\{d_1, d_2, ..., d_M, \sigma^2{}_M\}$ determined by the Yule–Walker equation (6.8). This calculation does not include the inverse Fourier transform but gives the following equations [6]:

$$c_y(0) = \ln \sigma^2_M$$

$$c_y(n) = -d_n - \sum_{k=1}^{n-1} \frac{k}{n} c(k) d_{n-k} \quad (1 \leqslant n \leqslant M)^{\cdot} \tag{6.15}$$

**Figure 6.6.** Comparison of $\log|X(\omega)|$ from FT and LP methods.

**Figure 6.7.** lPC-based cepstrum ($M = 380$).

Appendix A13.8 lists the EXCEL's VBA code created for this approach. For the voice sample shown in figure 6.4, the optimal order $M$ is 380, and we calculate the LPC-based cepstrum coefficients up to $M = 380$ terms, giving the following cepstrum (figure 6.7).

Figure 6.8 shows the real part of FT-based cepstrum up to 380 terms. Other than the vertical scale, two cepstra by FPC and FT exhibit the same structure.

As we have confirmed, cepstra calculated from the inverse Fourier transform of $\log|Y(\omega)|$ and linear prediction are equivalent. Both cepstra exhibit the same patterns and can be applied for speech analysis and voice recognition. The initial part ($c_g$) reflects the vocal cavity structure whereas the subsequent peaks are

**Figure 6.8.** FT-based cepstrum.

formants ($c_x$). Readers who are interested in cepstrum should refer to the references for detailed analysis.

Unfortunately, there is no muti-dimensional LPC (and ME) theory yet. Perhaps, one of the readers would conceive such theory!

## References

[1] Jain A K 1989 *Fundamentals of Digital Image Processing* (Englewood Cliffs, NJ: Prentice-Hall)

[2] Collomb C 2009 *Linear Prediction and Levinson-Durbin Algorithm* (http://emptyloop.com/technotes/A%20tutorial%20on%20linear%20prediction%20and%20Levinson-Durbin.pdf).

[3] Akaike H 1969 Fitting autoregressive models for prediction *Ann. Inst. Stat. Math* **21** 2. 437243E67
andAkaike H 1969 Power spectrum estimation through autoregressive model fitting *Ann. Inst. Stat. Math.* **21** 407–19

[4] Burg J 1967 Maximum entropy spectral analysis PhD Dissertation Department of Geophysics, Stanford University; and 1967 Maximum entropy spectral analysis *The 37th Annual Meeting, Soc. of Exploration Geophysicist*

[5] Cordes J 2015 *Lecture 17 Maximum Entropy Applications* (Ithaca, NY: Cornell University) www.astro.cornell.edu/~cordes/A6523/A6523_2015_Lecture17.pdf

[6] *LPC to/from Cepstral Coefficients* (MathWorks®) https://mathworks.com/help/dsp/ref/lpcto-fromcepstralcoefficients.html

# Fourier Transform and Its Applications Using Microsoft EXCEL® (Second Edition)

Shinil Cho

# Appendix A

## A1 Gibbs phenomena

The partial sum of the Fourier terms of the step function described in section 1.3.1 is given by

$$S_{2m+1} = \frac{4}{\pi}\left\{\sin x + \frac{1}{3}\sin 3x + \cdots + \frac{1}{2m+1}\sin(2m+1)\right\}.$$

Let us evaluate the value of $S_{2m+1}$ at the first peak by changing the partial sum into the following integral from

$$S_{2m+1}(x) = \frac{4}{\pi}\int_0^{\pi}\{\cos x + \cos 3x + \cdots + \cos(2m+1)x\}dx.$$

Because $2(\sin x)[\cos\{(2k+1)x\}] = \sin\{2(k+1)x\} - \sin(2kx)$, we obtain

$$S_{2m+1}(x) = \frac{2}{\pi}\int_0^x \frac{\sin[2(m+1)x]}{\sin x}dx = \frac{4}{\pi}\int_0^x D_{m+\frac{1}{2}}(2x)dx$$

where we define $D_n(x) = \frac{\sin[(n+1/2)x]}{2\sin(x/2)}$. $D_n(x)$ is an oscillating function of the $x$-axis, and thus $\int_0^x D_n(x)dx$ also oscillates, taking an extreme value at the zero point of $D_n(x)$. The first peak of $S_{2m+1}(x)$ is at $x = k_{m+1/2} = \pi/(2m+2)$ and

$$S_{2m+1}(x) = \frac{2}{\pi}\int_0^{\pi/(2m+2)} \frac{\sin[2(m+1)x]}{\sin x}dx = \frac{2}{\pi}\int_0^{\pi} \frac{\sin \xi}{\xi}\frac{\xi/(2m+2)}{\sin\{\xi/(2m+2)\}}d\xi.$$

Therefore,

$$\lim_{m\to\infty} S_{2m+1}(k_{m+1/2}) = \frac{2}{\pi}\int_0^{\pi} \frac{\sin \xi}{\xi}\left[\lim_{m\to\infty}\frac{\xi/(2m+2)}{\sin\{\xi/(2m+2)\}}\right]d\xi = \frac{2}{\pi}\int_0^{\pi} \frac{\sin \xi}{\xi}d\xi = 1.1789\ldots$$

## A2 Cauchy's residual theorem of the complex integral

Here is a short description of the Cauchy principle. Suppose a complex function $f(z)$ along a closed path C is regular on the region bounded by C except for a finite number of poles and isolated essential singularities in the interior of C. The theorem of residue states that the Cauchy principal value integral is given by

$$\oint_C f(z)dz = 2\pi i \sum (\text{residues})$$

where the sum is over all the poles and essential singularities inside C. If a pole lies at $x_0$ on the path C, the Cauchy principal value integral is given by

$$\oint f(z)dz = \pi i(\text{residues\_at\_} x_0) + 2\pi i \sum (\text{residues\_inside\_} C\}.$$

For a complete discussion of the Cauchy theorem and the complex integral, refer to appropriate mathematics books [1].

### A3 Fourier transform of a step function $U(\omega)$

Integral of equation (2.67)

$$u(t) = \frac{1}{2\pi} \int_{-\infty}^{\infty} U(\omega)e^{-i\omega t}d\omega = \frac{1}{2\pi}\lim_{\varepsilon \to 0^+} \int_{-\infty}^{\infty} \frac{e^{-i\omega t}}{i\omega - \varepsilon}d\omega$$

can be evaluated by using the Cauchy theorem.

(a) For t < 0, taking the complex integral path of figure A1(a), $u(t) = 0$.

(b) For t > 0, taking the integral path of figure A1(b), $u(t) = \frac{1}{2\pi}\lim_{\varepsilon \to 0^+} \int_{-\infty}^{\infty} \frac{e^{-i\omega t}}{i\omega - \varepsilon}d\omega = \frac{1}{2\pi i}\lim_{\varepsilon \to 0^+}[-2\pi i e^{-i(i\varepsilon)t}] = -1.$

Thus $u(t)$ is the step function: $u(t) = -1$ if $t > 0$; and 0 if $t < 0$.

### A4 Explicit form of the impulse response of forced harmonic oscillator

From equations (2.68) and (2.69),

$$g(t - t') = \frac{1}{2\pi} \int_{-\infty}^{\infty} \frac{1}{M(\omega_0^2 - \omega^2 - i\omega\Gamma)}e^{-i\omega(t-t')}d\omega$$

$$= -\frac{1}{2\pi M} \int_{-\infty}^{\infty} \frac{1}{(\omega^2 + i\omega\Gamma - \omega_0^2)}e^{-i\omega\tau}d\omega$$

(a) $t > 0$                      (b) $t < 0$

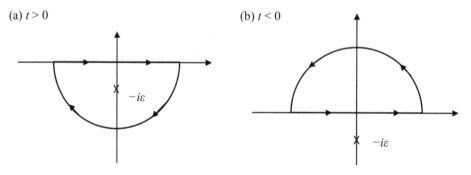

**Figure A1.** Integral paths of $u(t)$.

where $\tau = t - t'$. The roots of the denominator of the integrant are given by

$$\omega = \frac{-i\Gamma \pm \sqrt{4\omega_0^2 - \Gamma^2}}{2} \equiv \omega_\pm.$$

Because both $\omega_\pm$ are both on the lower half plane, $g(\tau) = 0$ for $\tau < 0$, reflecting the causality. For $\tau > 0$,

$$g(\tau) = -\frac{1}{2\pi M} \int_{-\infty}^{\infty} \frac{e^{-i\omega\tau}}{(\omega - \omega_+)(\omega - \omega_+)} d\omega = -\frac{1}{2\pi M} \frac{1}{\omega_+ - \omega_-} \int_{-\infty}^{\infty} \left( \frac{1}{\omega - \omega_+} - \frac{1}{\omega - \omega_-} \right) e^{-i\omega\tau} d\omega$$

$$= \frac{1}{2\pi M} \frac{1}{\omega_+ - \omega_-} 2\pi i [e^{-i\omega_+\tau} - e^{-i\omega_-\tau}] = \frac{i}{M} \frac{1}{\omega_+ - \omega_-} [e^{-i\omega_+\tau} - e^{-i\omega_-\tau}]$$

$$= \frac{i}{M} \frac{1}{\sqrt{4\omega_0^2 - \Gamma^2}} \left[ e^{-\frac{1}{2}\left(i\sqrt{4\omega_0^2 - \Gamma^2} + \Gamma\right)\tau} - e^{+\frac{1}{2}\left(i\sqrt{4\omega_0^2 - \Gamma^2} - \Gamma\right)\tau} \right].$$

Conversely, we can calculate the inverse Fourier transform of $g(t)$:

$$G(\omega) = \int_{-\infty}^{\infty} g(t) e^{i\omega t} dt = \frac{1}{M(\omega_0^2 - \omega^2 - i\omega\Gamma)}.$$

## A5 Hilbert transform and dispersion relation

Consider the following complex integral along the path showed in figure A2.

$$I = \oint \frac{f(\omega')d\omega'}{\omega' - \omega_r} \quad \text{where } \omega_r \text{ is real.}$$

By applying the Cauchy theorem, the line integral along the real axis should be the principal value integral defined symbolically as

$$P \int_{-\infty}^{\infty} = \int_{-\infty}^{\infty} \equiv \lim_{\varepsilon \to 0} \left[ \int_{-\infty}^{\omega_r - \varepsilon} + \int_{\omega_r + \varepsilon}^{\infty} \right],$$

and because there is no contribution from the semicircular path if $|f(\omega)| \to 0$ at infinity faster than $1/|\omega|$, the integral becomes

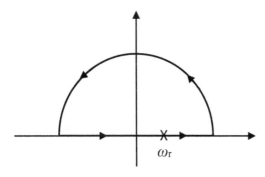

$\omega_r$

**Figure A2.** Integral path for the Hilbert transform.

$$I = \frac{1}{i\pi} \int_{-\infty}^{\infty} \frac{f(\omega')d\omega'}{\omega' - \omega_r}.$$

Separating the real and the imaginary parts, we obtain the Hilbert transform:

$$\text{Re}[f(\omega)] = \frac{1}{\pi} \int_{-\infty}^{\infty} \frac{\text{Im}[f(\omega')]d\omega'}{\omega' - \omega}$$

$$\text{Im}[f(\omega)] = -\frac{1}{\pi} \int_{-\infty}^{\infty} \frac{\text{Re}[f(\omega')]d\omega'}{\omega' - \omega}.$$

For simple cases where $\omega \geq 0$ and if $f(\omega)$ has 'even' or 'odd' symmetry,
(1) for the even symmetry, $f(-\omega) = f^*(\omega^*)$,

$$f(\omega) = \frac{1}{i\pi} \int_{-\infty}^{\infty} \frac{f(\omega')d\omega'}{\omega' - \omega} = \frac{1}{i\pi} \int_{\infty}^{0} \frac{f(-\omega')d\omega'}{\omega' + \omega} + \frac{1}{i\pi} \int_{0}^{\infty} \frac{f(\omega')d\omega'}{\omega' - \omega}$$

$$= \frac{1}{i\pi} \int_{\infty}^{0} \frac{f^*(\omega')d\omega'}{\omega' + \omega} + \frac{1}{i\pi} \int_{0}^{\infty} \frac{f(\omega')d\omega'}{\omega' - \omega}$$

$$= \frac{1}{i\pi} \int_{0}^{\infty} \frac{\omega(f(\omega') + f^*(\omega')) + \omega'(f(\omega') - f^*(\omega'))}{\omega'^2 - \omega^2}d\omega'$$

$$= \frac{1}{i\pi} \int_{0}^{\infty} \frac{2\omega \,\text{Re}[f(\omega')] + i2\omega' \,\text{Im}[f(\omega')]}{\omega'^2 - \omega^2}d\omega'$$

thus,

$$\text{Re}[f(\omega)] = \frac{2}{\pi} \int_{0}^{\infty} \frac{\omega' \,\text{Im}[f(\omega')]}{\omega'^2 - \omega^2}d\omega'$$

$$\text{Im}[f(\omega)] = -\frac{2}{\pi} \int_{0}^{\infty} \frac{\omega \,\text{Re}[f(\omega')]}{\omega'^2 - \omega^2}d\omega'.$$

this gives the dispersion relationship.
(2) For the odd symmetry, $f(-\omega) = -f^*(\omega^*)$,

$$f(\omega) = \frac{1}{i\pi} \int_{-\infty}^{\infty} \frac{f(\omega')d\omega'}{\omega' - \omega} = \frac{1}{i\pi} \int_{\infty}^{0} \frac{f(-\omega')d\omega'}{\omega' + \omega} + \frac{1}{i\pi} \int_{0}^{\infty} \frac{f(\omega')d\omega'}{\omega' - \omega}$$

$$= \frac{1}{i\pi} \int_{\infty}^{0} \frac{-f^*(\omega')d\omega'}{\omega' + \omega} + \frac{1}{i\pi} \int_{0}^{\infty} \frac{f(\omega')d\omega'}{\omega' - \omega}$$

$$= \frac{1}{i\pi} \int_{0}^{\infty} \frac{\omega(f(\omega') - f^*(\omega')) + \omega'(f(\omega') + f^*(\omega'))}{\omega'^2 - \omega^2}d\omega'$$

$$= \frac{1}{i\pi} \int_{0}^{\infty} \frac{2i\omega \,\text{Im}[f(\omega')] + 2\omega' \,\text{Re}[f(\omega')]}{\omega'^2 - \omega^2}d\omega'$$

thus,

$$\text{Re}[f(\omega)] = \frac{2}{\pi} \int_{0}^{\infty} \frac{\omega \,\text{Im}[f(\omega')]}{\omega'^2 - \omega^2}d\omega'$$

$$\text{Im}[f(\omega)] = -\frac{2}{\pi} \int_{0}^{\infty} \frac{\omega' \,\text{Re}[f(\omega')]}{\omega'^2 - \omega^2}d\omega'.$$

## A6 Derivation of dispersion relation $\chi''(\omega)$ and $\chi'(\omega)$

$$\chi(\omega) = \int_{0}^{\infty} g(\tau)e^{-i\omega\tau}d\tau$$

where $g(\tau) = \dfrac{i}{M}\dfrac{1}{\sqrt{4\omega_0^2 - \Gamma^2}}[e^{-\frac{1}{2}(\Gamma + i\sqrt{4\omega_0^2 - \Gamma^2})\tau} - e^{-\frac{1}{2}(\Gamma - i\sqrt{4\omega_0^2 - \Gamma^2})\tau}]$. Notice that the impulse response is real: $g^*(\tau) = g(\tau)$, and thus,

$$\chi_*(\omega) = \int_0^\infty g_*(\tau)e^{-i\omega\tau}d\tau = \int_0^\infty g(\tau)e^{i\omega(-\tau)}d\tau = \chi(-\omega).$$

From the Hilbert transform (2.76), we have

$$\frac{1}{\pi i}\int_{-\infty}^\infty \frac{\chi(\omega')}{\omega' - \omega}d\omega' = \frac{1}{\pi i}\int_{-\infty}^0 \frac{\chi(\omega')}{\omega' - \omega}d\omega' + \frac{1}{\pi i}\int_0^\infty \frac{\chi(\omega')}{\omega' - \omega}d\omega'$$

$$= \frac{1}{\pi i}\int_\infty^0 \frac{\chi(-\omega')}{\omega' + \omega}d\omega' + \frac{1}{\pi i}\int_0^\infty \frac{\chi(\omega')}{\omega' - \omega}d\omega' = \frac{1}{\pi i}\int_\infty^0 \frac{\chi_*(\omega')}{\omega' + \omega}d\omega' + \frac{1}{\pi i}\int_0^\infty \frac{\chi(\omega')}{\omega' - \omega}d\omega'$$

$$= \frac{1}{\pi i}\int_\infty^0 [\frac{\chi_*(\omega')}{\omega' + \omega} + \frac{\chi(\omega')}{\omega' - \omega}]d\omega' = \frac{1}{\pi i}\int_\infty^0 \frac{\omega\{\chi(\omega') + \chi_*(\omega')\} + \omega'\{\chi(\omega') - \chi_*(\omega')\}}{(\omega')^2 - \omega^2}d\omega'$$

$$= \frac{2}{\pi i}\int_\infty^0 \frac{\omega\, \mathrm{Re}[\chi(\omega')] + i\omega'\, \mathrm{Im}[\chi(\omega')]}{(\omega')^2 - \omega^2}d\omega'.$$

Thus, we obtain equation (2.75):

$$\chi'(\omega) = \frac{2}{\pi}\int_0^\infty \frac{\omega'\chi''(\omega')}{\omega'^2 - \omega^2}d\omega' \text{ and } \chi''(\omega) = -\frac{2}{\pi}\int_0^\infty \frac{\omega\chi'(\omega')}{\omega'^2 - \omega^2}d\omega'.$$

Using $\chi'(\omega)$ of equation (2.74) and replacing $\omega'$ with $z$,

$$\chi'(\omega) = \frac{2}{\pi}\int_0^\infty \frac{z}{z^2 - \omega^2}\frac{1}{M}\frac{z\Gamma}{(\omega_0^2 - z^2)^2 + z^2\Gamma^2}dz$$

$$= \frac{2\Gamma}{\pi M}\int_0^\infty \frac{z^2}{(z^2 - \omega^2)\{(\omega_0^2 - z^2)^2 + z^2\Gamma^2\}}dz \equiv \frac{\Gamma}{\pi M}\int_{-\infty}^\infty z^2 I(z)dz,$$

where $I(z) = \dfrac{z^2}{(z^2 - \omega^2)\{(\omega_0^2 - z^2)^2 + z^2\Gamma^2\}} = \dfrac{z^2}{(z^2 - \omega^2)(z^2 - \omega_0^2 - i\Gamma z)(z^2 - \omega_0^2 + i\Gamma z)}.$

Now, from $(z^2 - \omega_0^2 - i\Gamma z) = 0$, $z = \dfrac{i\Gamma \pm \sqrt{4\omega_0^2 - \Gamma^2}}{2} \equiv z_1, z_2$ and from $(z^2 - \omega_0^2 + i\Gamma z) = 0$, $z = \dfrac{-i\Gamma \pm \sqrt{4\omega_0^2 - \Gamma^2}}{2} \equiv z_3, z_4$, we obtain

$$I(z) = \frac{z^2}{(z + \omega)(z - \omega)(z - z_1)(z - z_2)(z - z_3)(z - z_4)}$$

$$\equiv \frac{A}{z - \omega} + \frac{B}{z + \omega} + \frac{C}{z - z_1} + \frac{D}{z - z_2} + \frac{E}{z - z_3} + \frac{F}{z - z_4}.$$

Because $z_3$ and $z_4$ are on the lower half plane, they do not contribute to the integral, and $\pm\omega$ are on the real axis. We now calculate residues at $\pm\omega$, $z_1$, and $z_2$ :

$$\mathrm{Res}(\omega) = \frac{\omega^2}{2\omega(\omega^2 - \omega_0^2 - i\Gamma\omega)(\omega^2 - \omega_0^2 + i\Gamma\omega)} = \frac{\omega}{2\{(\omega^2 - \omega_0^2) + \omega^2\Gamma^2\}},$$

$$\mathrm{Res}(-\omega) = -\frac{\omega}{2\{(\omega^2 - \omega_0^2) + \omega^2\Gamma^2\}}, \text{ and thus } \mathrm{Res}(\omega) + \mathrm{Res}(-\omega) = 0,$$

$$\mathrm{Res}(z_1) = \frac{z_1^2}{(z_1^2 - \omega^2)(z_1 - z_2)(z_1^2 - \omega_0^2 + i\Gamma z_1)} = \frac{z_1}{2i\Gamma\sqrt{4\omega_0^2 - \Gamma^2}(\omega_0^2 - \omega^2 + i\Gamma z_1)},$$

$$\mathrm{Res}(z_1) = -\frac{z_2}{2i\Gamma\sqrt{4\omega_0^2 - \Gamma^2}(\omega_0^2 - \omega^2 + i\Gamma z_2)}, \text{ and thus}$$

$$\text{Res}(z_1) + \text{Res}(z_2) = \frac{\omega_0^2 - \omega^2}{2i\Gamma\{(\omega_0^2 - \omega^2) + \omega^2\Gamma^2\}}.$$

Therefore,

$$\chi'(\omega) = \frac{\Gamma}{\pi M} \int_{-\infty}^{\infty} z^2 I(z) dz = \frac{\Gamma}{\pi M} \{2\pi i[\text{Res}(z_1) + \text{Res}(z_2)] + \pi i[\text{Res}(\omega) + \text{Res}(-\omega)]\}$$

$$= \frac{\Gamma}{M\pi} 2\pi i \frac{\omega_0^2 - \omega^2}{2i\Gamma\{(\omega_0^2 - \omega^2)^2 + \omega^2\Gamma^2\}} = \frac{1}{M} \frac{\omega_0^2 - \omega^2}{(\omega_0^2 - \omega^2)^2 + \omega^2\Gamma^2}.$$

Similarly,

$$\chi''(\omega) = -\frac{2}{\pi} \int_0^\infty \frac{\omega \chi'(\omega')}{\omega'^2 - \omega^2} d\omega' = -\frac{2}{\pi} \int_0^\infty \frac{\omega}{z^2 - \omega^2} \frac{1}{M} \frac{\omega_0^2 - z^2}{(\omega_0^2 - z^2)^2 + z^2\Gamma^2} dz$$

$$= -\frac{\omega}{\pi M} \int_{-\infty}^\infty I'(z) dz,$$

where $I'(z) = \dfrac{\omega_0^2 - z^2}{(z^2 - \omega^2)(\omega_0^2 - z^2)^2 + z^2\Gamma^2} = \dfrac{\omega_0^2 - z^2}{(z^2 - \omega^2)(z^2 - \omega_0^2 - i\Gamma z)(z^2 - \omega_0^2 + i\Gamma z)}$ .Residues are:

$$\text{Res}(z_1) = \frac{\omega_0^2 - z_1^2}{(z_1^2 - \omega^2)(z_1 - z_2)(z_1^2 - \omega_0^2 + i\Gamma z_1)}, \quad \text{and}$$

$$\text{Res}(z_2) = \frac{1}{2\sqrt{4\omega_0^2 - \Gamma^2}} \frac{1}{(\omega_0^2 - \omega^2 + i\Gamma z_2)}.$$

Also, $\text{Res}(\omega) + \text{Res}(-\omega) = 0$. Therefore, we obtain

$$\chi''(\omega) = -\frac{\omega}{\pi M} \int_{-\infty}^\infty I'(z) dz = -\frac{\omega}{\pi M} \{2\pi i[\text{Res}(z_1) + \text{Res}(z_2)] + \pi i[\text{Res}(\omega) + \text{Res}(-\omega)]\}$$

$$= -\frac{\omega}{\pi M} 2\pi i \frac{1}{2\sqrt{4\omega_0^2 - \Gamma^2}} \left\{ \frac{i\Gamma}{(\omega_0^2 - \omega^2 + i\Gamma z_2)} - \frac{i\Gamma}{(\omega_0^2 - \omega^2 + i\Gamma z_1)} \right\} = \frac{1}{M} \frac{\omega\Gamma}{(\omega_0^2 - \omega^2)^2 + \omega^2\Gamma^2}.$$

Q.E.D.

## A7 Winer–Khintchine theorem

The Wincr–Khintchine theorem states that if we define the power spectrum density $S(\omega)$

$$S(\omega) = \int_{-\infty}^{+\infty} R(\tau)e^{-i\omega\tau} d\tau,$$

then the correlation function $R(\tau)$ is given by

$$R(\tau) = \frac{1}{2\pi} \int_{-\infty}^{+\infty} S(\omega)e^{i\omega\tau} d\omega.$$

Proof: the power spectral density is defined as $S(\omega) = \lim_{T\to\infty} \dfrac{|X(\omega)|^2}{T}$. Suppose $x(t)$ is a signal data that has spectrum $X(\omega) = \int_{-\infty}^{+\infty} x(t)e^{-i\omega t} dt$, then

$$|X(\omega)|^2 = \left[\int_{-\infty}^{+\infty} x(t)e^{-i\omega t}dt\right]\left[\int_{-\infty}^{+\infty} x(\tau)e^{i\omega t}d\tau\right]$$

$$= \int_{-\infty}^{+\infty}\int_{-\infty}^{+\infty} e^{-i\omega(t-\tau)}x(t)x(\tau)dtd\tau = \int_{-\infty}^{+\infty} e^{-i\omega t'}\left[\int_{-\infty}^{+\infty} x(t'+\tau)x(\tau)d\tau\right]dt'.$$

Thus,

$$S(\omega) = \lim_{T\to\infty}\frac{|X(\omega)|^2}{T} = \int_{-\infty}^{+\infty} e^{-i\omega t'}[\lim_{T\to\infty}\frac{1}{T}\int_{-\infty}^{+\infty} x(t'+\tau)x(\tau)d\tau]dt' = \int_{-\infty}^{+\infty} e^{-i\omega t'}R(t')dt'$$

where $R(t') = \lim_{T\to\infty}\frac{1}{T}\int_{-\infty}^{+\infty} x(t'+\tau)x(\tau)d\tau$ is the correlation function. Therefore, the power spectral density $S(\omega)$ is given by the Fourier transform of correlation function $R(t')$. By taking the inverse Fourier transform, we obtain

$$R(\tau) = \frac{1}{2\pi}\int_{-\infty}^{+\infty} S(\omega)e^{i\omega\tau}d\omega.$$

## A8  Proof of equation (4.112)

For a stationary Gaussian modulation,

$$\phi(t) = \left\langle\exp\left[i\int_0^t \omega_1(\tau)d\tau\right]\right\rangle = \exp\left[-\langle\omega_1^2\rangle\int_0^t (t-\tau)\psi(\tau)d\tau\right].$$

Proof: define $\Phi(t) = \int_0^t \omega_1(\tau)d\tau$, then

$$\phi(t) = \langle\exp[i\Phi(t)]\rangle = 1 + \langle i\Phi(t)\rangle + \frac{1}{2!}\langle i\{\Phi(t)\}^2\rangle + \cdots$$

Because $\langle\omega_0\rangle = 0$ and $\langle\Phi(t)\rangle = 0$, and thus,

$$\phi(t) = \langle\exp[i\Phi(t)]\rangle = 1 - \frac{1}{2!}\langle\{\Phi(t)\}^2\rangle + \cdots$$

Comparing the above equation with the expansion of the following equation,

$$\exp[-\frac{1}{2}\langle\{\Phi(t)\}^2\rangle] = 1 - \frac{1}{2}\langle\{\Phi(t)\}^2\rangle + \cdots,$$

we use the following approximation:

$$\phi(t) = \left\langle\exp\left[-\frac{1}{2}\{\Phi(t)\}^2\right]\right\rangle = \exp\left(-\frac{1}{2}\int_0^t dt_1\int_0^t dt_2\langle\omega_1(t_1)\omega_1(t_2)\rangle\right).$$

The integral in the above equation, $I = \int_0^t dt_1\int_0^t dt_2\langle\omega_1(t_1)\omega_1(t_2)\rangle$, is carried out in the square region of $0\leqslant t_1\leqslant t$ and $0\leqslant t_2\leqslant t$ as shown in figure A3. Now, we divide the integral into two regions:

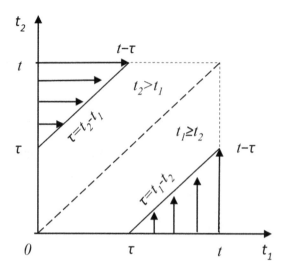

**Figure A3.** Integral regions of $\phi(t)$.

$I_1 = \int_0^t dt_1 \int_0^t dt_2 \langle \omega_1(t_1)\omega_1(t_2) \rangle$ where $t_1 \geq t_2$; and $I_2 = \int_0^t dt_1 \int_0^t dt_2 \langle \omega_1(t_1)\omega_1(t_2) \rangle$ where $t_1 \geq t_2$.

For a stationary Gaussian modulation, $\langle \omega_1(t_1)\omega_1(t_2) \rangle$ is an even function and depends on only $t_1 - t_2$. For the integral $I_1$ where $t_1 \geq t_2$, define $\tau = t_1 - t_2$, and then change the integral variables $\{t_2 \text{ and } t_1\}$ to $\{t_2 \text{ and } \tau\}$ and carry out the integrals with respect to $t_2$ and then $\tau$. The integral interval of $t_2$ becomes 0 to $t - \tau$ while $\langle \omega_1(t_1) \omega_1(t_2) \rangle$ is constant. Therefore, for $t_1 \geq t_2$,

$$I_1 = \int_0^t d\tau \int_0^{t-\tau} dt_2 \langle \omega_1(t_1)\omega_1(t_2) \rangle = \int_0^t (t - \tau)\langle \omega_1(0)\omega_1(\tau) \rangle d\tau.$$

For $t_2 > t_1$, define $\tau = t_2 - t_1$ change the integral variables to $\{t_1 \text{ and } \tau\}$ of integral $I_2$. By carrying out the integrals with respect to $t_1$ and then $\tau$, we obtain

$$I_2 = \int_0^t dt_1 \int_0^t dt_2 \langle \omega_1(t_1)\omega_1(t_2) \rangle = \int_0^t dt_2 \int_0^t dt_1 \langle \omega_1(t_1)\omega_1(t_2) \rangle$$
$$= \int_0^t d\tau \int_0^{t-\tau} dt_1 \langle \omega_1(t_1)\omega_1(t_2) \rangle = \int_0^t (t - \tau)\langle \omega_1(0)\omega_1(\tau) \rangle d\tau.$$

Therefore,

$$I = \int_0^t dt_1 \int_0^t dt_2 \langle \omega_1(t_1)\omega_1(t_2) \rangle = 2 \int_0^t (t - \tau)\langle \omega_1(0)\omega_1(\tau) \rangle d\tau = 2\langle \omega_1^2 \rangle \int_0^t (t - \tau)\psi(\tau) d\tau,$$

and we obtain

$$\phi(t) = \left\langle \exp\left[ -\frac{1}{2}\{\Phi(t)\}^2 \right] \right\rangle = \exp(-\langle \omega_1^2 \rangle) \int_0^t (t - \tau)\psi(\tau) d\tau$$

where we used equation (4.111) for a stationary Gaussian process.
    Q.E.D.

## A9 Fourier transform of Gaussian function

$$F(\omega) = \int_{-\infty}^{\infty} \exp\left[-\tfrac{1}{2}\varepsilon t^2\right] e^{-i\omega t} dt = e^{-\frac{\omega^2}{2\varepsilon}} \int_{-\infty}^{\infty} \exp\left[-\tfrac{1}{2}\varepsilon(t + \tfrac{i\omega}{\varepsilon})^2\right] dt$$

$$= e^{-\frac{\omega^2}{2\varepsilon}} \int_{-\infty+i\omega/\varepsilon}^{\infty+i\omega/\varepsilon} \exp\left[-\tfrac{1}{2}\varepsilon z^2\right] dz.$$

Consider the following complex integral $\int_{-\infty}^{\infty} \exp(-\varepsilon\frac{z^2}{2})dz$ where $z = x + iy$ along the close path as shown in figure A4. Because there is no singular point inside the integral path, the integral must be zero: $I_1 + I_2 + I_3 + I_4 = 0$ where each integral segment is:

$$I_1 = \int_{-R}^{R} \exp\left[-\tfrac{1}{2}\varepsilon x^2\right] dx, \quad I_2 = \int_{0}^{\omega'} \exp\left[-\tfrac{1}{2}\varepsilon(R + iy)^2\right] dy,$$

$$I_3 = \int_{R+i\omega'}^{-R+i\omega'} \exp\left[-\tfrac{1}{2}\varepsilon x^2\right] dx, \quad I_4 = \int_{\omega'}^{0} \exp\left[-\tfrac{1}{2}\varepsilon(-r + iy)^2\right] dy, \quad \text{and } \omega'' = \omega/\varepsilon.$$

Now, $\lim\limits_{R\to\infty} |I_2| \leqslant \lim\limits_{R\to\infty} \int_{0}^{\omega'} |\exp[-\frac{(R+iy)^2}{2}]| dy = \lim\limits_{R\to\infty} \int_{0}^{\omega'} \exp[-\frac{R^2}{2}]\exp[\frac{y^2}{2}] dy = 0$, and

$$\lim\limits_{R\to\infty} |I_4| \leqslant \lim\limits_{R\to\infty} \int_{0}^{\omega''} \exp\left[-\frac{R^2}{2}\right]\exp\left[\frac{y^2}{2}\right] dy = 0.$$

Thus, from the complex integral along the closed loop, $\lim\limits_{R\to\infty} I_1 = -\lim\limits_{R\to\infty} I_3$, i.e.

$$-\lim\limits_{R\to\infty} I_3 = \int_{-\infty+i\omega/\varepsilon}^{\infty+i\omega/\varepsilon} \exp\left[-\tfrac{1}{2}\varepsilon x^2\right] dx = \lim\limits_{R\to\infty} I_1 = \int_{-\infty}^{\infty} \exp\left[-\tfrac{1}{2}\varepsilon x^2\right] dx = \sqrt{\frac{2\pi}{\varepsilon}}.$$

Therefore, $F(\omega) = \int_{-\infty}^{\infty} \exp[-\tfrac{1}{2}\varepsilon t^2] e^{-i\omega t} dt = \sqrt{\frac{2\pi}{\varepsilon}}$.

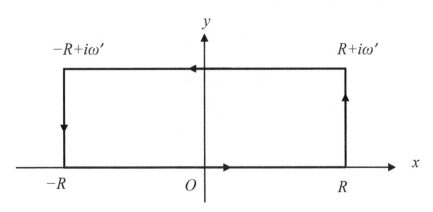

**Figure A4.** Integral path of Gaussian integral.

## A10  Fourier transform of $\exp(\pm\lambda r)/r$

$$I = \int \frac{e^{\pm i\lambda r}}{r} e^{-i\vec{k}\cdot\vec{r}} d\vec{r} = \int_0^\infty r^2 dr \int_0^\pi \sin\theta d\theta \int_0^{2\pi} d\phi \frac{e^{\pm i\lambda r}}{r} e^{-ikr\cos\theta} = 2\pi \int_0^\infty r^2 dr \int_0^\pi \sin\theta d\theta \frac{e^{\pm i\lambda r}}{r} e^{-ikr\cos\theta}$$

$$= 2\pi \int_0^\infty r^2 dr \int_{-1}^{+1} d\xi \frac{e^{\pm i\lambda r}}{r} e^{-ikr\xi} (\xi = \cos\theta)$$

$$= 2\pi \int_0^\infty r^2 dr \frac{e^{\pm i\lambda r}}{r} \frac{e^{-ikr} - e^{ikr}}{-ikr} = \frac{2\pi}{ik} \int_0^\infty [e^{i(\pm\lambda+k)r} - e^{i(\pm\lambda ik)r}] dr.$$

this integral does not converge but we may multiply a factor $e^{-\varepsilon r}$ ($\varepsilon > 0$) and let $\varepsilon \to 0$ after carrying out the integral:

$$I = \frac{2\pi}{ik} \lim_{\varepsilon\to 0} \int_0^\infty (e^{i[\pm\lambda r - kr] - \varepsilon r} - e^{i[\pm\lambda r + kr] - \varepsilon r}) dr$$

$$= \frac{1}{4\pi^2} \frac{1}{ik} \lim_{\varepsilon\to 0} \int_0^\infty (e^{-[\varepsilon - i(\pm\lambda) + ik]r} - e^{-[\varepsilon - i(\pm\lambda) - ik)]r}) dr$$

$$= 2\pi \frac{1}{k^2 - \lambda^2}.$$

## A11  Partial integral in the uncertainty calculation

$$\text{Integral } (\Delta x)^2 = \int_{-\infty}^{+\infty} \left| \frac{dF(k)}{dk} \right|^2 dk.$$

Proof:

$$(\Delta x)^2 = <x^2> = \int_{-\infty}^{+\infty} x^2 |f(x)|^2 dx = \int x^2 f^*(x) \left( \int F(k) e^{ikx} dk \right) dx$$

$$= \int_{-\infty}^{+\infty} f^*(x) \left( \int_{-\infty}^{+\infty} F(k) \left( -\frac{d^2 e^{ikx}}{dk^2} \right) dk \right) dx.$$

By the partial integral, we obtain

$$\int_{-\infty}^{+\infty} F(k) (\frac{d^2 e^{ikx}}{dk^2}) dk = \left[ F(k) \frac{de^{ikx}}{dk} \right]_{-\infty}^{+\infty} - \int_{-\infty}^{+\infty} \frac{dF(k)}{dk} \cdot \frac{de^{ikx}}{dk} dk$$

$$= - \left[ \frac{dF(k)}{dk} e^{ikx} \right]_{-\infty}^{+\infty} + \int_{-\infty}^{+\infty} \frac{d^2 F(k)}{dk} e^{ikx} dk.$$

The first term vanishes for localized wave packets. Thus,

$$\int_{-\infty}^{+\infty} F(k)(\frac{d^2 e^{ikx}}{dk^2}) dk = - \int_{-\infty}^{+\infty} \frac{dF(k)}{dk} \cdot \frac{de^{ikx}}{dk} dk$$

$$= - \left[ \frac{dF(k)}{dk} e^{ikx} \right]_{-\infty}^{+\infty} + \int_{-\infty}^{+\infty} \frac{d^2 F(k)}{dk} e^{ikx} dk = \int_{-\infty}^{+\infty} \frac{d^2 F(k)}{dk} e^{ikx} dk.$$

Therefore, we obtain

$$(\Delta x)^2 = \int_{-\infty}^{+\infty} f*(x)(\int_{-\infty}^{+\infty} F(k)(-\frac{d^2 e^{ikx}}{dk^2})dk)dx = -\int_{-\infty}^{\infty} f*(x)(\int_{-\infty}^{+\infty} \frac{d^2 F(k)}{dk}e^{ikx}dk)dx$$

$$= -\int_{-\infty}^{+\infty}(\int_{-\infty}^{+\infty} f*(x)e^{ikx}dx)\frac{d^2 F(k)}{dk^2}dk = -\int_{-\infty}^{+\infty} F*(k)\frac{d^2 F(k)}{dk^2}dk$$

$$= -\left[ F*(x)\frac{dF(k)}{dk} \right]_{-\infty}^{+\infty} + \int_{-\infty}^{+\infty} \frac{dF*(k)}{dk} \cdot \frac{dF(k)}{dk}dk = \int_{-\infty}^{+\infty}\left| \frac{dF(k)}{dk} \right|^2 dk.$$

## A12 Notes on EXCEL

### A12.1 Autofill

*Autofill* is a very useful EXCEL capability that is frequently used for scientific calculations. With the capability, iterative calculations can be carried out without programming. Here is a simple example that enters integer 0, 1, 2, 3, .... in column A (figure A5).

 (1) Input 0 in cell A1.
 (2) Go to cell A2 and enter = **A1 + 1** and press <Enter>. The value of cell A2 becomes 1.
 (3) Place the cursor in cell A2 and hit <Enter>. The cell is emphasized.
 (4) Click on the fill handle (a small solid square at the lower right corner). The cursor becomes a plus sign (+). While pressing the right mouse button, drag the cursor to cells below in column A to until reaching to the desired integer value.

Remark: there is a quicker method when a formula is involved. Providing that there is data in a column adjacent to the one with the formula, all that is required is to right double-click the fill handle and the formula will automatically be copied down the column [2]

### A12.2 Adding 'data analysis'

Because EXCEL's default setting is not for scientific computation, we need to add data analysis tools in order to add the Fourier transform option and other useful

**Figure A5.** Example of autofill.

scientific calculation tools [3]. From the [File] menu, select [Options] to display the 'EXCEL option' screen. Click on [Add-Ins] to display the following screen (figure A6).

Next, click on [Go] to display available add-ins, and then check [Analysis ToolPak], [Analysis ToolPak - VBA], and [Solver Add-In] (figure A7).

Remark: [Solver] is an optimization routine used for applications such as linear programming.

### A12.3 Enabling EXCEL macro

Excel macro is a visual basic programing environment. Appendix A13.4 lists the source codes created for this book. Those who are interested in Excel macro, refer to *Excel® VBA for Physicists* of this Concise Series [4].

Take the following steps to enable EXCEL's macro capability:

(1) Go to [Trust Center] (figure A8).
(2) From [Option], go to [Trust Center], and click on [Trust Center Settings...] of [Excel Options] in [Microsoft Excel Trust Center].
(3) Select [Macro Settings] and check [Enable All Macro (not recommended; potentially dangerous code can run)] (figure A9). Click [OK] to complete the setting.
(4) In the pulldown menu, click on [View] and click on [Macros] to select [View Macros] and create a macro (visual basic) program (figure A10).

**Figure A6.** EXCEL options screen.

**Figure A7.** Available add-ins.

**Figure A8.** Excel options.

**Figure A9.** Macro settings.

**Figure A10.** Macro entry.

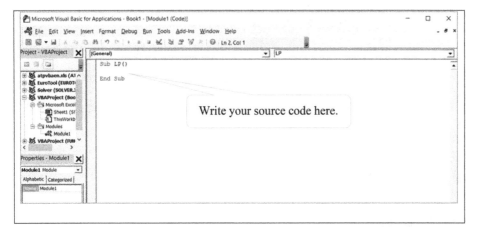

**Figure A11.** Macro editor.

After entering a macro name, we can create its source code using a built-in editor (figure A11).

# A13 Lists of VBA codes and EXCEL files for figures

## A13.1 Fourier series of square pulse train

```
Sub FTSeries()
Cells(1,1)="Fourier series of square pulse train of height =1 and period = 2π"
N = 100 'Number of terms
    For j = 1 To 512
        temp = 0
        For k = 1 To N
                temp = temp + Sin(0.02 * (2 * k - 1) * j) / (2 * k - 1)
        Next k
                Cells(j, 2) = temp * 4 / 3.1416
    Next j
End Sub
```

## A13.2 Macro for rearranging terms

```
Sub Arrange()
        'This macro reads number data (odd and even terms) from Column B
        'and copy odd terms to Column C,
        'and even terms to Column D.
        'There are 1024 data points starting cell B3 in Column B.
N=1024 'Number of data terms
        For k = 1 To N
            a = Cells(2 * k, 2)
                Cells(k + 1, 3) = a
            b = Cells(2 * k + 1, 2)
                Cells(k + 1, 4) = b
        Next k
End Sub
```

### A13.3 Drawing 3D surface chart of $z = f(x, y)$

**Figure 1.11:** A mode $u_{3,2}$ of the two-dimensional standing wave on a membrane given by equation (1.28)

$$u_{3,2}(x, y) = \sin(3\pi x)\sin(2\pi y) \text{ with } a = b = 3.15$$

can be drawn using EXCEL's 3D surface chart in the following steps (figure A12).

*x-axis:*
- (1) Enter **0** in Cell C2.
- (2) Enter = **C2 + 0.105** in Cell D2.
- (3) *Autofill* to Cell AG (to the right).

*y-axis:*
- (4) Enter **0** in Cell B3.
- (5) Enter = **B3 + 0.5** in Cell B4.
- (6) *Autofill* to Cell B33 (downward).

*Enter the function:*
- (7) Click on Cell C3 and enter = **SIN(3*$B3)*SIN(2*C$2)** in Cell C3.
- (8) *Autofill* to Cell AG (to the right).
- (9) Continue *autofill* to B33 (downward) to complete the computation.

*3D Surface Graph to make the diffraction pattern:*
- (10) Highlight Cell C3 to AG33.
- (11) Follow [Insert] → [Charts] → [All Charts] → [3-D Surface] to draw $u_{3,2}$.
- (12) Enlarge the graph for a better view.

*3D Rotation:*
- (13) Right-click on the 3D surface to display a popup window and select [3-D Rotation]. The [Format Chart Area] menu is displayed on the right side of the screen. Select [3-D Rotation] item. Enter appropriate angle values for a best view of figure 1.11.

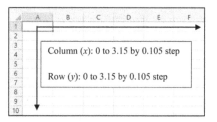

**Figure A12.** Preparing an EXCEL sheet for a dipole field.

**Figure 4.34**: For the diffraction pattern through a rectangular hole given by equation (4.58) with $C' = 1$, $D_x = 1$, $D_y = 1$, and $k/2R = 1$,

$$U_P(x_s, y_s) = \frac{\sin(x_s)}{x_s} \frac{\sin(y_s)}{y_s},$$

we use discrete points of $-15 \leqslant x \leqslant +15$, $-15 \leqslant y \leqslant +15$, with an increment of 0.4. The increment is selected to avoid the singular point $(0, 0)$.

*x-axis:*
    (14) Enter **15** in Cell C2.
    (15) Enter = **C2 + 0.4** in Cell D2 (to the right).
    (16) *Autofill* to Cell BZ2 (downward).

*y-axis:*
    (17) Enter **15** in Cell B3.
    (18) Enter = **B3 + 0.5** in Cell B4.
    (19) *Autofill* to Cell BZ78.

*Diffraction pattern:*
    (20) Click on Cell C3 and enter = **((SIN($B3)/$B3)*(SIN(C$2)/C$2))^2**.
    (21) *Autofill* to Cell BZ3 (to the right).
    (22) Continue *autofill* to BZ78 (downward) to complete the computation.

*3D Surface Graph to make the diffraction pattern:*
    (23) Highlight Cell B2 to BZ78.
    (24) Follow [Insert] → [Charts] → [All Charts] → [3-D Surface] to draw $U_P(x_s, y_s)$.
    (25) Enlarge the graph for a better view.

*3D Rotation:*
    (26) Right-click on the 3D surface to display a popup window and select [3-D Rotation]. The [Format Chart Area] menu is displayed on the right side of the screen. Select [3-D Rotation] item. Enter **0°** in the [X Rotation] box and **90°** in the [Y Rotation] box to display the top view of the chart as shown in figure 4.34.

**Figure 4.36:** The diffraction pattern through a circular hole given by equation (4.59) with $2\pi C' = 1$, $D = 2$, and $k/R = 1$:

$$U_P(x_s, y_s) = \frac{J_1(\rho_s)}{(\rho_s)}.$$

In order to draw figure 4.36,
    (27) Start from step (14).

(28) In step (20), enter **= (BESSELJ(SQRT($B3^2 + C$2^2),1)/SQRT ($B3^2 + C$2^2))^2** in Cell B3 where BESSELJ(x,n) is the EXCEL's Bessel function $J_n(x)$.

## A13.4 VBA code and example of 1D-FFT

```
Sub FFT1()
Cells(1, 1) = "1D-FFT" 'f=0.1Hz-sine wave plus 0.3Hz sine wave
'Reserved 8196 data points.
Dim x(8192)            'Raw data. Can be complex data.
Dim ReData(8192)       'Real part of raw data {x(i); i=0,2,..,N-1}.
Dim ImData(8192)       'Imaginary part of raw data {x(i); i=0,2,..,N-1}.
Dim ReFFT(8192)        'Real part of data after bit reverse.
Dim ImFFT(8192)        'Imaginary part of data after bit reverse.
Dim Co(8192)           'cosine-value.
Dim Sn(8192)           'sine-value.
    N = 512            '# of data points.
    NN = N - 1
    Sampling = 1 'Sampling frequency (Hz).
    Res = Sampling / N ' Resolution of FFT in Hz.
    M = Int(Log(N) / Log(2) + 0.5) 'N=2^M.
    FTop = 1 'FFT. If FTop =-1 then inverse FFT.
    Pi = 3.14159
    Cells(2, 1) = "N=": Cells(2, 2) = N
    Cells(2, 4) = "N=2^M": Cells(2, 5) = "M=": Cells(2, 6) = M
    Cells(2, 7) = "Sampl.period": Cells(2, 8) = Sampling: Cells(2, 9) = "sec"
    Cells(2, 10) = "Freq res=": Cells(2, 11) = Res: Cells(2, 12) = "Hz"
    Cells(3, 1) = "Data #"
    Cells(3, 2) = "x(i)"
    Cells(3, 4) = "F-step(Hz)"
    Cells(3, 5) = "Re[FFT]"
    Cells(3, 6) = "Im[FFT]"
    Cells(3, 7) = "ABS[FFT]"
'The data number is shown from Cell (4,1) in Column A.
'Raw data {x(i)} are given from Cell(4,2) in Column B.
'This program reads the existing raw data from spread sheet.
For i = 0 To NN
    x(i) = Cells(4 + i, 2) 'Read raw data {x(i); i=0,2,..,N-1}.
        ReData(i) = WorksheetFunction.ImReal(x(i))        'Real part of data
        ImData(i) = WorksheetFunction.Imaginary(x(i))     'Imaginary part of data
Next i
            GoSub BitReverse
            GoSub FFT1
'Write FFTed signal on spread sheet.
    For i = 0 To NN
        Cells(4 + i, 4) = i * Freq 'Frequency step in Column D
        Cells(4 + i, 5) = ReFFT(i) 'Real part of FFT in Column E
        Cells(4 + i, 6) = ImFFT(i) 'Imaginary part of FFT in Column F
        Cells(4 + i, 7) = Sqr(ReFFT(i) ^ 2 + ImFFT(i) ^ 2) 'Intensity in Column G
    Next i
Exit Sub
'******************************************************
BitReverse:
'Bit reversion.
'I1=bit string to be revered.
'I2= Reversed bit of I1.
For i = 0 To NN
    I1 = i: I2 = 0: I3 = 1
            For k = 1 To M
            I1 = I1 + I1
                If I1 >= N Then I2 = I2 + I3: I1 = I1 - N   'MSB =1
                    I3 = I3 + I3
        Next k
            ReFFT(I2) = ReData(i)
```

```
                  ImFFT(I2) = ImData(i)
     Next i
Return
'******************************************************
FFT1:
'Butterly compuation of 1-dim FFT.
J1 = N / 2
J2 = J1 - 1
K1 = 1: K2 = K1 - 1
For i = 0 To J2
    W = -2 * FTop * Pi * i / N 'Index of exponential term; w=-2*pai*i/N
        Co(i) = Cos(W)
        Sn(i) = Sin(W)
Next i
For L0 = 1 To M
    For L1 = 0 To J2
        For L2 = 0 To K2
                A0 = 2 * K1 * L1 + L2
                A1 = A0 + K1
                W0 = J1 * L2
                W1 = ReFFT(A1) * Co(W0) - ImFFT(A1) * Sn(W0)
                W2 = ReFFT(A1) * Sn(W0) + ImFFT(A1) * Co(W0)
                ReFFT(A1) = ReFFT(A0) - W1
                ImFFT(A1) = ImFFT(A0) - W2
                ReFFT(A0) = ReFFT(A0) + W1
                ImFFT(A0) = ImFFT(A0) + W2
            Next L2
    Next L1
        J1 = J1 / 2
        J2 = J1 - 1
        K1 = K1 * 2
        K2 = K1 - 1
Next L0
    If FTop = 1 Then GoTo skip
        For i = 0 To NN
            ReFFT(i) = ReFFT(i) / N
            ImFFT(i) = ImFFT(i) / N
        Next i
skip: Return
'******************************************************
End Sub
```

**Example:** FFT of $y = \sin(0.2\pi t) + 2\sin(0.3\pi t)$; $\Delta t = 1$, $N = 512$ and thus $\Delta f = 1/N\Delta t = 1.95 \times 10^{-3}$ Hz (figure A13).

**Figure A13.** An example of one-dimensional FFT using the VBA code.

A-19

## A13.5 VBA code and example of 2D-FFT

```
Sub FFT2()
Cells(1, 1) = "512x512 2-Dimensional FFT (space-wave number)"
Dim xy(512, 512)      'Raw data. Note: it will be transposed matrix of the 1st FFT.
Dim ReDataTemp(512)  'Real part of raw data (xy(i,j); i, j=0,2,..,N-1}.
Dim ImDataTemp(512)  'Imaginary part of raw data ({xy(i, j); i, j=0,2,..,N-1}.
Dim FFT1(512, 512)    'Output from FFT.
Dim ReFFTtemp(512)    'Temporal real-values for FFT.
Dim ImFFTtemp(512)    'Temporal imaginary-values for FFT.
Dim FFT1xTemp(512, 512) 'Real part of FFT1(i, j).
Dim FFT1yTemp(512, 512) 'Imaginary part of FFT1(i, j).
        'NOTE: After the 1st FFT, xy(i,j)=Transposed(FFTxTemp(i, j), FFTyTemp(i, j)).
Dim FFTxTemp(512, 512)  'Transposed matrix element of the 2nd FFT.
Dim FFTyTemp(512, 512)  'Transposed matrix element of the 2nd FFT.
Dim FFT(512, 512)       'FFT(i,j)=Transposed (FFT1xTemp(i, j}, FFT1yTemp(i,j))}).
Dim Co(512)            'cosine-value.
Dim Sn(512)            'sine-value.
    N = 512            '# data points.
    NN = N - 1
    d = 1              'Two-dimensional square lattice spacing.
    Res = 1 / N * d    'Resolution of FFT.
    M = Int(Log(N) / Log(2) + 0.5)        'N=2^M.
    FTop = 1           'FFT. If FTop =-1 then inverse FFT.
    Pi = 3.14159
    Cells(2, 1) = "N=": Cells(2, 2) = N    'N by N lattice.
    Cells(2, 4) = "N=2^M": Cells(2, 5) = "M=": Cells(2, 6) = M
    Cells(2, 7) = "Lattice space": Cells(2, 8) = d
    Cells(2, 10) = "Resolution=": Cells(2, 11) = Res    'FFT resolution
'Raw data xy(i,j) are computed below.
    For i = 0 To NN
        For j = 0 To NN
            xy(j, i) = Sin(0.2 * Pi * i) * Sin(0.2 * Pi * j) + 2 * Sin(0.4 * Pi
* i) * Sin(0.8 * Pi * j)
        Next j
    Next i
'*****************************************************************************
****
'First FFT to get FFT1(i, j):[X']=[W][x]
    GoSub FFT2
'Transpose FFT1-matrix to get [X']^t=[Wx]^t=([x]^t)[W] because[W]^t=[W].
    For i = 0 To NN
        For j = 0 To NN
            FFTxTemp(i, j) = WorksheetFunction.ImReal(FFT1(i, j))
            FFTyTemp(i, j) = WorksheetFunction.Imaginary(FFT1(i, j))
            xy(j, i) = WorksheetFunction.Complex(FFTxTemp(i, j), FFTyTemp(i, j))
        Next j
    Next i
'Second FFT to get [X]^t=[W]([x]^t)[W]=[W][X']
    GoSub FFT2
'Transpose FFT1-matrix [X]^t to get FFT(i,j): [X]=([X]^t)^t.
    For i = 0 To NN
        For j = 0 To NN
            FFTxTemp(i, j) = WorksheetFunction.ImReal(FFT1(i, j))
            FFTyTemp(i, j) = WorksheetFunction.Imaginary(FFT1(i, j))
            FFT(j, i) = WorksheetFunction.Complex(FFTxTemp(i, j), FFTyTemp(i, j))
        Next j
    Next i
'Compute and output ABS(FFT(i,j))
For j = 0 To N / 2 - 1
```

```
        For i = 0 To N / 2 - 1
            Cells(9 + i, 5 + j) = WorksheetFunction.ImAbs(FFT(i, j))   'Intensity of
2D-FFT
        Next i
    Next j
    'Labeling axes and steps for output data
        Cells(9, 3) = "FFTx(column)"
        Cells(8, 5) = "FFTy(row)"
            For i = 0 To N / 2 - 1
                Cells(9 + i, 4) = i * Res
            Next i
            For j = 0 To N / 2 - 1
                Cells(9, 5 + j) = j * Res
            Next j
    Exit Sub
    '*******************************************************************************
    FFT2:
    'Bit reversion and 1D-FFT for a given j (the x-components of matrix).
    For j = 0 To NN              'x-component
        For i = 0 To NN          'y-component
            ReDataTemp(i) = WorksheetFunction.ImReal(xy(i, j))
            ImDataTemp(i) = WorksheetFunction.Imaginary(xy(i, j))
        Next i
                        GoSub BitReversion
        '1D-FFT using bit-reversed data {ReFFTtemp(I2)and ImFFTtemp(I2)}
        'Output from FFT is {ReFFTtemp(I2)and ImFFTtemp(I2)}
                        GoSub FFT1
        'Get the matrix element (i,j) of FFT1(i,j)=[W][x] for a given j.
        For i = 0 To NN
            FFT1xTemp(i, j) = ReFFTtemp(i)
            FFT1yTemp(i, j) = ImFFTtemp(i)
            FFT1(i, j) = WorksheetFunction.Complex(FFT1xTemp(i, j), FFT1yTemp(i, j))
    'Transposition
        Next i
    Next j
    Return
    '***************************************************
    BitReversion:
    'Bit reversion for a given j (the x-components of matrix).
    'I1=bit string to be revered.
    'I2= Reversed bit of I1.
    For i = 0 To NN
        I1 = i: I2 = 0: I3 = 1
            For k = 1 To M
                I1 = I1 + I1
                    If I1 >= N Then I2 = I2 + I3: I1 = I1 - N 'Corresponding to MSB =1.
                        I3 = I3 + I3
            Next k
                ReFFTtemp(I2) = ReDataTemp(i)
                ImFFTtemp(I2) = ImDataTemp(i)
    Next i
    Return
    '***************************************************
    FFT1:
    'Butterly computation of 1D-FFT for a given j (the x-components of matrix).
    j1 = N / 2
    j2 = j1 - 1
    k1 = 1: k2 = k1 - 1
```

```
For i = 0 To j2
    W = -2 * FTop * Pi * i / N 'Index of exponential term; w=-2*pai*i/N.
        Co(i) = Cos(W)
        Sn(i) = Sin(W)
Next i
For L0 = 1 To M
    For L1 = 0 To j2
        For L2 = 0 To k2
                A0 = 2 * k1 * L1 + L2
                A1 = A0 + k1
                W0 = j1 * L2
                W1 = ReFFTtemp(A1) * Co(W0) - ImFFTtemp(A1) * Sn(W0)
                W2 = ReFFTtemp(A1) * Sn(W0) + ImFFTtemp(A1) * Co(W0)
                ReFFTtemp(A1) = ReFFTtemp(A0) - W1
                ImFFTtemp(A1) = ImFFTtemp(A0) - W2
                ReFFTtemp(A0) = ReFFTtemp(A0) + W1
                ImFFTtemp(A0) = ImFFTtemp(A0) + W2
            Next L2
        Next L1
            j1 = j1 / 2
            j2 = j1 - 1
            k1 = k1 * 2
            k2 = k1 - 1
Next L0
    If FTop = 1 Then GoTo Skip
        For i = 0 To NN
            ReFFTtemp(i) = ReFFTtemp(i) / N
            ImFFTtemp(i) = ImFFTtemp(i) / N
        Next i
Skip: Return
'****************************************************************************
End Sub
```

**Example:** 2D (512×512)-point FFT. $\Delta t = 1$, $N = 512$ and thus $\Delta f = 1/N\Delta t = 1.95 \times 10^{-3}$ Hz.

The displayed image has $256 \times 256$ pixels (figure A14).

2D-FFT of $f(x, y) = \sin(0.2\pi x)\sin(0.2\pi y) + 2\sin(0.4\pi x)\sin(0.8\pi y)$

**Figure A14.** An example of two-dimensional FFT using the VBA code.

A-22

## A13.6 VBA code of quantum walk

```
Sub Qwalk()
Cells(1, 1) = "Quantum walk"
    N = 2001 'N=odd integer and the center is at (N+1)/2.
    Dim Phi1(4000)
    Dim Phi2(4000)
    Dim Phi1temp(4000)
    Dim Phi2temp(4000)
        U11 = 0.707107: U12 = U11: U21 = U11: U22 = -U11     'H-gate as Coin matrix
        P11 = U11: P12 = U12: P21 = 0: P22 = 0               'Shift matrix P
        Q11 = 0: Q12 = 0: Q21 = U21: Q22 = U22               'Shift matrix Q
'Initial condition
t = 0   'Dummy time & time increment=1
    For i = 0 To N + 2
        Phi1temp(i) = 0
        Phi2temp(i) = 0
    Next i
' Center lattice point [1,0]^t at N/2+1
        Phi1temp((N + 1) / 2) = 1
        Phi2temp((N + 1) / 2) = 0
'------------------------------------------
Repeat:
t = t + 1
    For i = (N + 1) / 2 To N
```

```
        Phi1(i) = P11 * Phi1temp(i + 1) + P12 * Phi2temp(i + 1) + Q11 *
Phi1temp(i - 1) + Q12 * Phi2temp(i - 1)
        Phi2(i) = P21 * Phi1temp(i + 1) + P22 * Phi2temp(i + 1) + Q21 *
Phi1temp(i - 1) + Q22 * Phi2temp(i - 1)
    Next i
    For i = (N - 1) / 2 To 1 Step -1
        Phi1(i) = P11 * Phi1temp(i + 1) + P12 * Phi2temp(i + 1) + Q11 *
Phi1temp(i - 1) + Q12 * Phi2temp(i - 1)
        Phi2(i) = P21 * Phi1temp(i + 1) + P22 * Phi2temp(i + 1) + Q21 *
Phi1temp(i - 1) + Q22 * Phi2temp(i - 1)
    Next i
        For j = 0 To N + 2
            Phi1temp(j) = Phi1(j)
            Phi2temp(j) = Phi2(j)
        Next j
If t < 1000 Then GoTo Repeat
    Cells(1, 3) = "# positions": Cells(1, 5) = N
    Cells(1, 2) = "Time="
    Cells(1, 3) = t
    Cells(3, 1) = "Position"
    Cells(3, 2) = "Phi(x)"
    Cells(3, 3) = "Phi(y)"
    Cells(3, 4) = "|Phi(x,y)|"
        For i = 1 To N
            Cells(i + 3, 1) = i
            Cells(i + 3, 2) = Phi1temp(i)
            Cells(i + 3, 3) = Phi2temp(i)
            Cells(i + 3, 4) = Sqr(Phi1temp(i) ^ 2 + Phi2temp(i) ^ 2)
        Next i
End Sub
```

### A13.7 Levinson algorithm

Suppose we have the Yule–Walker equation [4–8]

$$R(k) + \sum_{n=1}^{M} d_n R(k-n) = \sigma_M^2 \text{ if } k = 0;$$

$$= 0 \text{ if } 1 \leqslant k \leqslant M$$

where we used the order of terms $M$ and the standard deviation is denoted as $\sigma_M^2$. The Levinson algorithm finds the linera prediction coefficients recursively. In the matrix form, the Yule–Walker equation (where $M$ is replaced with $n$) is given by

$$\mathbb{R}_n \vec{d}_n = \vec{v}_n$$

where

$$\mathbb{R}_n = \begin{bmatrix} R(0)R(1)\cdots\cdots R(n) \\ R(1)R(0)\cdots\cdots R(n-1) \\ \vdots \ddots \vdots \\ R(n)R(n-1)\cdots R(0) \end{bmatrix}, \ \vec{d}_M = \begin{bmatrix} 1 \\ d_1 \\ \vdots \\ d_n \end{bmatrix}, \text{ and } \vec{v}_M = \begin{bmatrix} \sigma_n^2 \\ 0 \\ \vdots \\ 0 \end{bmatrix}.$$

When we increase $n$ by 1, i.e. $n$ to $n+1$, we need to solve $\mathbb{R}_{n+1}\vec{d}_{n+1} = \vec{v}_{n+1}$, i.e.

$$\mathbb{R}_{M+1} = \begin{bmatrix} R(0)R(1)\cdots\cdots R(n+1) \\ R(1)R(0)\cdots\cdots R(n) \\ \vdots \ddots \vdots \\ \vdots \ddots \vdots \\ R(n+1)R(n)\cdots R(0) \end{bmatrix} = \begin{bmatrix} & R(n+1) & \\ & \vdots & \\ \mathbb{R}_n & \vdots & \\ & R(n+1)R(n)\cdots R(0) & \end{bmatrix},$$

$$\vec{d}'_{n+1} = \begin{bmatrix} 1 \\ d'_1 \\ \vdots \\ \vdots \\ d'_{n+1} \end{bmatrix}, \text{ and } \vec{v}_{n+1} = \begin{bmatrix} \sigma_{n+1}^2 \\ 0 \\ \vdots \\ 0 \end{bmatrix}.$$

In order to calculate $\vec{d}'_{n+1}$, the Levinson algorithm uses previously calculated $\vec{d}_n$. Consider the following equation:

$$\begin{bmatrix} R(0)R(1)\cdots\cdots R(n)R(n+1) \\ R(1)R(0)\cdots\cdots R(n-1)R(n) \\ \vdots \ddots \vdots \quad \vdots \vdots \\ \vdots \ddots \vdots \quad \vdots \vdots \\ R(n)R(n-1)\cdots\cdots R(0)R(1) \\ R(n+1)R(n)\cdots\cdots R(1)R(0) \end{bmatrix} \left\{ \begin{bmatrix} 1 \\ d_1 \\ \vdots \\ \vdots \\ d_n \\ 0 \end{bmatrix} + c_{M+1} \begin{bmatrix} 0 \\ d_n \\ \vdots \\ \vdots \\ d_1 \\ 1 \end{bmatrix} \right\} = \left\{ \begin{bmatrix} \sigma_n^2 \\ 0 \\ \vdots \\ \vdots \\ 0 \\ \Delta_{n+1} \end{bmatrix} + c_{M+1} \begin{bmatrix} \Delta_{n+1} \\ 0 \\ \vdots \\ \vdots \\ 0 \\ \sigma_n^2 \end{bmatrix} \right\}.$$

Therefore, if we define

$$\vec{d}'_{n+1} = \begin{bmatrix} 1 \\ d'_1 \\ \vdots \\ \vdots \\ \vdots \\ d'_{n+1} \end{bmatrix} = \begin{bmatrix} 1 \\ d_1 \\ \vdots \\ \vdots \\ d_n \\ 0 \end{bmatrix} + c_{M+1} \begin{bmatrix} 0 \\ d_n \\ \vdots \\ \vdots \\ d_1 \\ 1 \end{bmatrix} \text{ and } \sigma_{n+1}^2 = \sigma_n^2 + c_{n+1}\Delta_{n+1} = \sigma_n^2(1 - c_{n+1}^2),$$

then $\mathbb{R}_{n+1}\vec{d}_{n+1} = \vec{v}_{n+1}$ can be calculated. Therefore, starting with $n = 0$, and increasing $n$ by 1 until it reaches to $M$, we can repeat the above alogrithm to find $\{d_1, d_2, \ldots., d_M, \sigma^2{}_M\}$. From the above equation, we obtain $\Delta_{n+1} = \sum\limits_{k=0}^{n} R_{n+1-k}d_k$, where $d_0 = 1$.

Let $c_{n+1}$ such that $\Delta_{n+1} + c_{n+1}\sigma_n^2 = 0$, i.e., $c_{n+1} = -\Delta_{n+1}/\sigma_n^2$, then

$$\begin{bmatrix} R(0) \cdots\cdots R(n+1) \\ R(0) \\ \vdots \ddots \quad \vdots \\ \vdots \quad \ddots \quad \vdots \\ \vdots \quad \ddots \quad \vdots \\ R(n+1)\cdots\cdots R(0) \end{bmatrix} \left\{ \begin{bmatrix} 1 \\ d_1 \\ \vdots \\ \vdots \\ d_n \\ 0 \end{bmatrix} + c_{M+1}\begin{bmatrix} 0 \\ d_n \\ \vdots \\ \vdots \\ d_1 \\ 1 \end{bmatrix} \right\} = \begin{bmatrix} \sigma_n^2 + c_{n+1}\Delta_{n+1} \\ 0 \\ \vdots \\ 0 \\ 0 \end{bmatrix}.$$

Therefore, if we define

$$\vec{d}'_{n+1} = \begin{bmatrix} 1 \\ d'_1 \\ \vdots \\ d'_{n+1} \end{bmatrix} = \begin{bmatrix} 1 \\ d_1 \\ \vdots \\ d_n \\ 0 \end{bmatrix} + c_{M+1}\begin{bmatrix} 0 \\ d_n \\ \vdots \\ d_1 \\ 1 \end{bmatrix} \quad \text{and} \quad \sigma_{n+1}^2 = \sigma_n^2 + c_{n+1}\Delta_{n+1} = \sigma_n^2(1 - c_{n+1}^2),$$

then $\mathbb{R}_{n+1}\vec{d}_{n+1} = \vec{v}_{n+1}$ can be calculated. Therefore, starting with $n = 0$, and increasing $n$ by 1 until it reaches $M$, we can repeat the above algorithm to find $\{d_1, d_2, \ldots., d_M, \sigma^2{}_M\}$.

**Algorithm flow:**

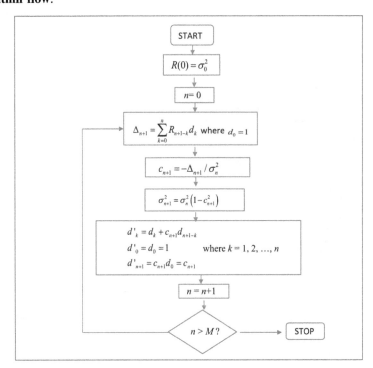

## A13.8 Linear prediction coefficients and power spectrum

```
Sub LinearPrediction()
Dim NMAX As Integer
Dim MMAX As Integer
Dim d() As Double      'LPC coefficients
Dim data() As Double 'Raw data
Dim x() As Double      'Estimated data
Dim xms As Double      'Mean square discrepancy
Dim wk1()
Dim wk2()
Dim wkm()
    NMAX = 4096 'Max data array for EXCEL
    MMAX = NMAX 'Max order of fitting terms
    'M <NMAX
    'N = number of raw data points
ReDim data(NMAX)
ReDim x(NMAX)
ReDim d(MMAX)
ReDim wk1(NMAX)
ReDim wk2(NMAX)
ReDim wkm(MMAX)
'Given a real time series signal data set data(n), and given order M,
'this routine returns linear prediction coefficients as d(M),
'and returns the mean square discrepancy as xms.
'Read data array from column B.
Samplfreq = 44100  'Assume sampling frequency = 44.10 kHz.
N = 1024       'N < NMAX
M = 185        'M < MMAX
    Cells(1, 5) = "Terms: N="
    Cells(2, 5) = N
    Cells(3, 5) = "Order: M="
    Cells(4, 5) = M
'Read raw data from Column B.
For j = 1 To N
    data(j) = Cells(j, 2)
Next j
    sumx = 0     'DC offset.
        For j = 1 To N
            sumx = sumx + data(j)
        Next j
        avx = sumx / N
            For k = 1 To N
                data(k) = data(k) - avx
            Next k
'Levinson algorithm starts here.
    p = 0
        For j = 1 To N
            p = p + data(j) ^ 2
        Next j
    xms = p / N
    wk1(1) = data(1)
    wk2(N - 1) = data(N)
        For j = 2 To N - 1
            wk1(j) = data(j)
            wk2(j - 1) = data(j)
        Next j
            For k = 1 To M
                wkm(k) = 0
            Next k
```

```
For k = 1 To M
        pneum = 0
        denom = 0
            For j = 1 To N - k
                pneum = pneum + wk1(j) * wk2(j)
                denom = denom + wk1(j) ^ 2 + wk2(j) ^ 2
            Next j
                d(k) = 2 * pneum / denom
                xms = xms * (1 - d(k) ^ 2)
                    For i = 1 To k - 1
                        d(i) = wkm(i) - d(k) * wkm(k - i)
                    Next i
'       The Levinson algorithm is recursive, building up the answer
'       for larger and lager values of M until the desired value is reached.
'       At this point in the algorithm, one could return the coefficient
'       vector d and scalar xms for a set of LP coefficients
'       with k (rather than M) terms.
'       Definition: dt=sampling interval in the time domain.
            If k = M Then GoTo Skip
                For i = 1 To k
                    wkm(i) = d(i)
                Next i
                    For j = 1 To N - k - 1
                        wk1(j) = wk1(j) - wkm(k) * wk2(j)
                        wk2(j) = wk2(j + 1) - wkm(k) * wk1(j + 1)
                    Next j
Next k
'Output d(M) to column D for the given N
'Cell E1 = xms
        Cells(1,5)=xms
Skip:
    For j = 1 To M
        Cells(j, 4) = d(j)
    Next j
'Calculation of LPC values

    For i = 1 To M
        x(i) = data(i)
        Cells(i, 6) = x(i)
    Next i
'Estimated function x(i)
'Calculating FPE and AIC values
    errvalue = 0
    Cells(7, 5) = "Error value="
    Cells(9, 5) = "FPC="
    Cells(10, 5) = "AIC="
        For i = M + 1 To N
            x(i) = 0
                For j = 1 To M
                    x(i) = x(i) + d(j) * data(i - j)
                Next j
            Cells(i, 6) = x(i)
            errvalue = errvalue + (data(i) - x(i)) ^ 2
        Next i
            errvalue = errvalue / (N - M)
            Cells(8, 5) = errvalue
                FPE = (N + M + 1) / (N - M - 1) * errvalue
                Cells(10, 5) = FPE
```

```
                          AIC = N * Log(errvalue) + 2 * M
                          Cells(11, 5) = AIC
'Power spectrum using Max Entropy
'Given d, m, xms as retuned by the above,
'this function returns the power spectrum estimate P(f) as a
'function of fdt.
'data = N data points and sampling with Samplfreq (Hz) in this example
'Freq step = Samplfreq/N (=44100/4096=10.77) Hz
 For i = 1 To N
    dt = 1 / Samplfreq 'sampling time interval
       'total time = n*dt=0.09288 sec = 1/frequency step
         freq = 10.77 * (i - 1)
       'frequency values i=0, 1, 2, ...., N-1
         Cells(i, 8) = freq
         fdt = freq * dt
         theta = 6.28318530717959 * fdt       'Angular freq=2*pai*fdt
         sumr = 1
         sumi = 0
         wr = 1
         wi = 0
             For k = 1 To M
                 wtemp = wr
                 wr = wr * Cos(theta) - wi * Sin(theta)
                 wi = wi * Cos(theta) + wtemp * Sin(theta)
                 sumr = sumr - d(k) * wr
                 sumi = sumi - d(k) * wi
             Next k
       Power = xms / (sumr ^ 2 + sumi ^ 2)  ' Power spectrum
       Cells(i, 9) = Power
   Next i
End Sub
```

## A13.9  Calculation of cepstrum coefficients from linear prediction method

```
Sub LPC_Cepstrum()
'Calculation of LPC Cepstrum,
'assuming the LPC {d(n)} are obtained and stored in Column D,
'and the mean square discrepancy, xms, is stored in cell E1.
'xms is sigma.
Dim NMAX As Integer
Dim MMAX As Integer
Dim d() As Double     ' LPC coefficients
Dim cc() As Double    ' Cepstrum coefficients
Dim xms As Double     ' Mean square discrepancy
    NMAX = 8196 ' max data array for EXCEL
    MMAX = NMAX ' max order of fitting terms
    'M <MMAX
    'N = number of raw data points
ReDim d(MMAX)
ReDim cc(MMAX)
'Read xms from Cell E1 and d(n) from Column D
          xms = Cells (1,5)
      For j = 1 To M
         d(j)=Cells(j, 4)
      Next j
           cconst = Log(xms)
           Cells(1, 6) = cconst
             For k = 1 To M
                 cc(k) = 0
             Next k
               cc(1) = d(1)
             For k = 2 To M
                   For j = 1 To k - 1
                       cc(k) = cc(k) + (j / k) * cc(j) * d(k - j)
                   Next j
                     cc(k) = cc(k) + d(k)
                     Cells(k, 6) = cc(k)
            Next k
End Sub
```

# References

[1] Riley K F, Hobson M P and Bence S J 2002 *Mathematical Methods for Physics and Engineering* (Cambridge: Cambridge University Press)
[2] www.excel-easy.com/examples/autofill.html
[3] Bloch S C 2003 *EXCEL for Engineers and Scientists* (New York: Wiley)
[4] Liengme B V 2016 *EXCEL® VBA for Physicist A Primer IOP Concise Physics* (San Rafael, CA: Morgan & Claypool)

Printed in the USA
CPSIA information can be obtained
at www.ICGtesting.com
JSHW061315120124
55268JS00004B/44